Stephen Moore

Computational 3D Modeling of Hemodynamics in the Circle of Willis

Stephen Moore

Computational 3D Modeling of Hemodynamics in the Circle of Willis

VDM Verlag Dr. Müller

Impressum/Imprint (nur für Deutschland/ only for Germany)

Bibliografische Information der Deutschen Nationalbibliothek: Die Deutsche Nationalbibliothek verzeichnet diese Publikation in der Deutschen Nationalbibliografie; detaillierte bibliografische Daten sind im Internet über http://dnb.d-nb.de abrufbar.

Alle in diesem Buch genannten Marken und Produktnamen unterliegen warenzeichen-, marken- oder patentrechtlichem Schutz bzw. sind Warenzeichen oder eingetragene Warenzeichen der jeweiligen Inhaber. Die Wiedergabe von Marken, Produktnamen, Gebrauchsnamen, Handelsnamen, Warenbezeichnungen u.s.w. in diesem Werk berechtigt auch ohne besondere Kennzeichnung nicht zu der Annahme, dass solche Namen im Sinne der Warenzeichen- und Markenschutzgesetzgebung als frei zu betrachten wären und daher von jedermann benutzt werden dürften.

Coverbild: www.purestockx.com

Verlag: VDM Verlag Dr. Müller Aktiengesellschaft & Co. KG
Dudweiler Landstr. 99, 66123 Saarbrücken, Deutschland
Telefon +49 681 9100-698, Telefax +49 681 9100-988, Email: info@vdm-verlag.de
Zugl.: Christchurch, University of Canterbury, 2006

Herstellung in Deutschland:
Schaltungsdienst Lange o.H.G., Berlin
Books on Demand GmbH, Norderstedt
Reha GmbH, Saarbrücken
Amazon Distribution GmbH, Leipzig
ISBN: 978-3-639-10399-1

Imprint (only for USA, GB)

Bibliographic information published by the Deutsche Nationalbibliothek: The Deutsche Nationalbibliothek lists this publication in the Deutsche Nationalbibliografie; detailed bibliographic data are available in the Internet at http://dnb.d-nb.de.

Any brand names and product names mentioned in this book are subject to trademark, brand or patent protection and are trademarks or registered trademarks of their respective holders. The use of brand names, product names, common names, trade names, product descriptions etc. even without a particular marking in this works is in no way to be construed to mean that such names may be regarded as unrestricted in respect of trademark and brand protection legislation and could thus be used by anyone.

Cover image: www.purestockx.com

Publisher:
VDM Verlag Dr. Müller Aktiengesellschaft & Co. KG
Dudweiler Landstr. 99, 66123 Saarbrücken, Germany
Phone +49 681 9100-698, Fax +49 681 9100-988, Email: info@vdm-publishing.com
Copyright © 2008 VDM Verlag Dr. Müller Aktiengesellschaft & Co. KG and licensors
All rights reserved. Saarbrücken 2008

Printed in the U.S.A.
Printed in the U.K. by (see last page)
ISBN: 978-3-639-10399-1

Computational 3D Modelling of Hemodynamics in the Circle of Willis

Stephen Moore

A thesis presented for the degree of
Doctor of Philosophy
in
Mechanical Engineering
at the
University of Canterbury,
Christchurch, New Zealand.

2 July 2008

Acknowledgements

First of all I would like to thank my supervisor and friend Professor Tim David, both for the opportunity to undertake this research and for the opportunities you have given me to participate in related projects, collaborating with various research groups. Your guidance and support has helped make this four years the most intellectually stimulating period of my life so far.

Secondly I would like to thank Dr John Fink for the significant amount of time you have spent filling in the large gaps in my anatomical and physiological knowledge, as well as making sure that our work maintains a clinical focus, so that hopefully one day it will find its way into the medical field.

I would like to thank Dr Richard Watts for the time you have spent sharing your vast knowledge of MRI with me, in order to answer every annoyingly minute detail that I trouble over.

Finally I would like to thank my wonderful partner and soul mate Nikal, who painstakingly went through every single page of this thesis with a fine tooth comb, to turn my sloppy writing style into english. That's when you know it's love.

Contents

Acknowledgements   iii

Nomenclature   xxxiii

Abstract   xxxix

1 Anatomy   **1**
  1.1 Introduction   1
  1.2 Cerebral Arteries   2
  1.3 Collateral Circulation   9
  1.4 Variability of the circle of Willis 14
  1.5 Blood Vessel Structure 19

2 Physiology   **23**
  2.1 Introduction 23
  2.2 Brain Metabolism 23
  2.3 Blood 26
  2.4 Smooth Muscle Contraction 34
  2.5 Clinical Mechanics of Blood Flow 37
  2.6 Mechanisms for the Regulation of Blood Flow 40

3 Stroke   **53**
  3.1 Introduction 53
  3.2 Hemorrhagic Stroke 54
  3.3 Ischemic Stroke 55

4 Literature Review   **63**
  4.1 Introduction 63
  4.2 Circle of Willis Modelling 64
    4.2.1 1D Models 65
    4.2.2 2D Models 69
    4.2.3 3D Models 70
  4.3 Cerebral Autoregulation Modelling 72

5 Geometry Generation   **83**
  5.1 Introduction 83
  5.2 Idealized Geometry 87
  5.3 Patient Specific Geometry 95

6 Cerebral Autoregulation Modelling   **117**
  6.1 Introduction 117
  6.2 Idealized Circle of Willis Autoregulation Model 120
  6.3 Patient Specific Circle of Willis Autoregulation Model 124

7 Fluid Dynamics Modelling   **145**
  7.1 Introduction 145
  7.2 Governing Equations and Discretization 145
  7.3 Blood Viscosity Models 150
  7.4 Boundary Conditions 152
  7.5 Simulation Procedure 158

8 Idealized Results   **161**
  8.1 Introduction 161
  8.2 Complete Circle of Willis 162
  8.3 Fetal P1 Circle of Willis 166
  8.4 Missing A1 170

9 Patient Specific Results   **177**
  9.1 Introduction 177
    9.1.1 Mesh Convergence 177
    9.1.2 Pulsatile Effects 180
    9.1.3 Non Newtonian Effects 184
  9.2 Complete Circle of Willis 187
    9.2.1 LICA Occlusion 188
    9.2.2 RICA Occlusion 192
    9.2.3 BA Occlusion 197
  9.3 Missing Posterior Communicating Artery 200
    9.3.1 LICA Occlusion 200
    9.3.2 RICA Occlusion 204
    9.3.3 BA Occlusion 209
  9.4 Fused Anterior Cerebral Arteries 213
    9.4.1 LICA Occlusion 213
    9.4.2 RICA Occlusion 217
    9.4.3 BA Occlusion 220

10 Conclusions and Future Work   **227**

A Magnetic Resonance Imaging   **235**

   A.1 Introduction 235

   A.2 Nuclear Magnetic Characteristics of Elements 235

   A.3 Generation and Detection of the Magnetic Resonance Signal 239

   A.4 Localization of the Magnetic Resonance Signal 247

   A.5 Data Acquisition and Image Reconstruction 251

   A.6 Angiography 258

      A.6.1 Time Of Flight 258

      A.6.2 Phase Contrast 263

B Marching Cubes   **271**

C Fluid Dynamics   **283**

   C.1 Introduction 283

   C.2 Governing Equations 283

      C.2.1 Reynolds Transport Theorem 286

      C.2.2 Conservation of Mass and Momentum 290

   C.3 Discretization 299

      C.3.1 Evaluation of the Convective Term 306

      C.3.2 Evaluation of the Diffusive Term 308

      C.3.3 Evaluation of the Source Term 313

      C.3.4 Evaluation of the Unsteady Term 314

      C.3.5 Pressure Velocity Coupling 315

      C.3.6 Monitoring Convergence 322

      C.3.7 Solver Procedure 323

      C.3.8 Boundary Conditions 324

      C.3.9 Multigrid Methods 325

List of Figures

1.1   A description of terminology useful in describing structures within the
      body, while an individual is in the anatomical position.   2

1.2   An inferior transverse view of the brain, illustrating the afferent and effer-
      ent arteries visible from this viewpoint.   4

1.3   A midsagittal view of the brain, illustrating the afferent and efferent ar-
      teries visible from this viewpoint.   5

1.4   A sagittal view of the brain, illustrating the afferent and efferent arteries
      visible from this viewpoint.   6

1.5   A schematic of the cerebral arteries, labelled by their abbreviations and
      illustrating the general regions of the brain which they supply with blood.   8

1.6   The cerebral arteries comprising and immediately surrounding the circle
      of Willis.   9

1.7   Collateral parallel pathway for blood flow given an occlusion of the ICA. . 10

1.8   Collateral parallel pathway for blood flow given an occlusion of the BA. . 11

1.9   Collateral transverse anastomoses for blood flow via the circle of Willis,
      given an occlusion of the ICA. 12

1.10 Collateral transverse anastomoses between the major cerebral arteries,
      given an occlusion of the MCA. 13

1.11 Collateral parallel and transverse anastomoses for blood flow, given an occlusion of the CCA. 13

1.12 Anatomic configurations in the anterior circle of Willis found in the study by Hartkamp et al [58]. 15

1.13 Anatomic configurations in the posterior circle of Willis found in the study by Hartkamp et al [58]. 16

1.14 The generalized structure of the layers of the blood vessel wall, observable in most vessels except for the capillaries and venules. 20

2.1 The overall processes involved in carbohydrate metabolism as occurs in the cerebral tissue. Aerobic respiration involves the breakdown of glucose into pyruvic acid in the process of glycolysis, pyruvic acid inters the tricarboxylic acid cycle within the mitochondrion of the cell, producing high energy molecules NADH and FADH$_2$ which can be used to synthesize the majority of ATP in the electron transport chain. Anaerobic respiration involves the breakdown of pyruvic acid into lactic acid and a much smaller amount of ATP. 25

2.2 The hemoglobin molecule contains four globular protein chains, each tightly coupled with a heme group. The heme group is a porphyrin ring with an iron atom at its centre to which oxygen can bind. In addition other products of metabolism such as carbon dioxide and subsequently hydrogen ions, can bind to the protein chains. 28

2.3 (a) The oxyhemoglobin saturation curve, illustrating the variation in the amount of oxygen that binds to hemoglobin as a function of the oxygen tension, and also the effects of other substances on the position of the curve (b) the actual concentration of oxygen in the blood stream (including that dissolved in the plasma) as a function of oxygen tension. Note: the pink dashed line illustrates the normal oxygen tension in arterial blood. 30

2.4 (a) the carbamino saturation curve illustrating the variation in the amount of carbon dioxide that binds to hemoglobin as a function of the carbon dioxide tension, and also the effects of other substances on the position of the curve (b) the actual concentration of carbon dioxide in the blood stream (including that dissolved in the plasma and that in the form of bicarbonate) as a function of carbon dioxide tension. Note: the pink dashed line illustrates the normal carbon dioxide tension in arterial blood. 32

2.5 The overall processes involved in delivery of oxygen to and the removal of carbon dioxide from the cerebral tissue. 33

2.6 A schematic of a single smooth muscle cell, illustrating the important Actin and Myosin myofilaments and the effect if a contraction on the shape of the cell. 35

2.7 The physiological processes involved in a smooth muscle contraction. Note: the major regulator is intracellular calcium which either enters the cell through channels in the plasma membrane or is released from internal stores. 36

2.8 The baroreceptor reflex involves a feedback from the receptors at the aortic arch and carotid sinus, to the medulla oblongata, which in turn alters heart rate and stroke volume as well as blood vessel tone. Note: the baroreceptor reflex has maximal sensitivity about small deviations from the normal blood pressure, but adapts this normal level over a period of days. 41

2.9 The Renin-Angiotensin-Aldosterone mechanism regulates blood pressure by the alteration of blood volume and level of blood vessel constriction. The mechanism acts in the long term via the release of hormones from the kidney and liver. 43

2.10 The Antidiuretic Hormone mechanism regulates blood pressure via the alteration of blood volume and the level of blood vessel constriction. The mechanism acts in the long term via the release of hormones from the pituitary gland in the brain. 43

2.11 A standard cerebral autoregulation curve, illustrating the relatively constant level of blood flow over a range of cerebral perfusion pressures. The autoregulated region involves the vascular smooth muscle altering its tension and hence the caliber of the blood vessels, thereby altering the overall cerebrovascular resistance to blood flow. Outside of the autoregulated region (i.e. at very high or low blood pressures) where the smooth muscle is maximally constricted or dilated respectively, the blood flow is then a function of the blood pressure. 45

2.12 The more commonly proposed mechanisms involved in the various components of the cerebral autoregulation mechanism. Note: the $\oplus$ symbol indicates a pathway which promotes an increase in muscle tone and the $\ominus$ symbol indicates a pathway which promotes a decrease in muscle tone. . . . 47

3.1 Hemorrhagic stroke occurs when blood from a burst artery is forced into either the brain tissue, or into the narrow space between the brain surface and the layer of tissue that covers it. 54

3.2 Ischaemic stroke occurs when an artery in the brain becomes blocked and the supply of blood to a region of brain tissue is cut off. 55

4.1 A schematic of a 1D network model of the circle of Willis. Black represents the main arteries of the circle of Willis (incorporated in all works). Purple illustrates the anterolateral central arteries (ALA) and the anteromedial and posteromedial anastomoses (AM and PM) included in the work of Hudetz et al [67] (note: the VA's were not included in this work). Pink illustrates the common and external carotid arteries (CCA and ECA) and the periorbital anastomosis included in the work of Viedma et al [164]. Green illustrates a simplification of the arterioles, capillaries, venules and veins included in the work of Zagzoule et al [168]. 66

4.2 A schematic of the 2D model of the circle of Willis employed in the work of David et al [36, 45]. 69

4.3 The MRI segmented 3D circle of model used in the work of Cebral et al [31]. 71

4.4   The MRI segmented 3D circle of Willis model used in the work of Kim et
      al [80].   72

4.5   (a) The 2D model of the cerebral tissue and blood vessels used in the work
      of Ursino et al [156] and (b) the division of the cerebral vasculature used
      in the subsequent work [157].   74

4.6   A schematic of the autoregulation model implemented in the work of
      Ursino et al [153], illustrating the various components of the autoregu-
      lation mechanism and its relationship between blood flow, pressure and
      smooth muscle tone.   75

4.7   A schematic illustrating the electrical network analogy implemented in the
      work of Ursino et al [155].   76

4.8   A schematic illustrating the cardiovascular model with additional physi-
      ological control systems and their relationships, implemented in work of
      Ursino et al [96].   77

4.9   A schematic of the cerebral vascular compartments and relationships im-
      plemented in the patient simulator autoregulation model of Thoman et al
      [144].   79

4.10 A schematic of the physiological processes involved in the autoregulatory
      response in the work of Banaji et al [16].   80

4.11 A schematic of the five compartments and the associated chemical path-
      ways implemented in the autoregulation model of Banaji et al [16]. Orange
      represents the vascular lumen compartment and the associated blood bio-
      chemistry, pink represents the vascular smooth muscle compartment and
      the reactions involving actin and myosin myofilaments in the production
      of force, light blue, yellow and green represent the extracellular, intracel-
      lular and mitochondrial parts of the cerebral tissue respectively, involving
      a series of reactions related to metabolism as well as the movement of ions
      and other metabolites.   82

5.1   (a) A transverse *maximum intensity projection (MIP)* of the cerebral vas-
      culature, obtained from a 3D time of flight scan and (b) a sagittal view of a
      DSA scan of the arteries perfused by a single ICA, showing much greater
      detail of the cerebral vasculature. Note: the definition of a maximum
      intensity projection is given in Appendix A.   85

5.2   The procedure for creating the skeletal structure for the idealized models.
      (a) three orthogonal MR projections, (b) splines traced over the arterial
      segments, (c) the projection of any two orthogonal splines to create a
      projected curve of an artery or piece of an artery and (d) the complete
      skeletal structure.   88

5.3   (a) The projected curve created into a 3D sketch with points then placed
      along the arterial segments, (b) circular cross sections sketched on planes
      placed along the 3D sketch and (c) the surface loft of the circular cross
      sections to create the arterial wall.   89

5.4   The procedure for creating junctions when the parent and daughter arteries
      are of similar diameter and small bifurcation angles. (a) crescent shaped
      surface lofts, (b) filling the surface and (c) the completed junction.   90

5.5   The procedure for creating junctions when the parent and daughter ar-
      teries are of greatly differing diameter and large bifurcation angles. (a)
      An extruded circular surface used to trim the parent artery, (b) the edge
      of the hole lofted to the end of the smaller daughter artery and (c) the
      completed junction.   90

5.6   A complete idealized 3D model of a normal complete circle of Willis.   91

5.7   Axial views of the (a) Fetal P1 and (b) Missing A1, idealized circle of Willis
      anatomical variations created. Note: the models have been shown trun-
      cated part way along the internal carotid and basilar arteries to provide a
      clearer view of the efferent and communicating arteries.   92

5.8   An example of the tetrahedral element mesh created for the idealized
      circle of Willis models.   94

5.9   The basis for marching cubes illustrated with (a) a stack of MRI images
      from a 3D TOF scan forming the 3D array M(x,y,z), (b) two consecutive
      slices in the array with the 4 vertices used per slice highlighted, (c) the 8
      array values used to form the vertices of the cube and (d) and example
      triangulation through a cube, where light blue and dark blue represent
      MR intensity values which are lower and higher than the isosurface value,
      or vice versa. 96

5.10 Problems associated with directly segmenting the circle of Willis surface
      topology from the MRI data. (a) the presence of unwanted blood vessels
      and other tissues in the dataset causing (b) unwanted surface topology
      in the segmented model (c) the frequent poor visibility of the posterior
      communicating arteries. 97

5.11 (a) Lumps in the arterial wall surface and coarseness resulting from the
      limited resolution and noise in the MRI scan, (b) the result of smoothing
      the dataset to remove the lumps and (c) the result of interpolating the
      dataset to achieve a finer surface. 98

5.12 A transverse MIP of the circle of Willis illustrating (a) a rectangular ROI
      drawn around the circle of Willis, (b) the dataset reduced to the rectangu-
      lar ROI and (c) the dataset subsequently interpolated, effectively dividing
      up the voxels present in the original dataset. 99

5.13 A coronal MIP of the circle of Willis illustrating (a) a rectangular ROI
      drawn around the circle of Willis, (b) the dataset reduced to the rectangu-
      lar ROI and (c) the dataset subsequently interpolated, effectively dividing
      up the voxels present in the original dataset. 100

5.14 A transverse MIP illustrating the effect of a dynamically altering isosurface
      value (a) a value too low to segment all of the required geometry (b) an
      isosurface value which is too low and causes a significant amount of noise
      to be segmented but also places the arterial wall in the wrong location and
      (c) an isosurface value which is appropriate to segment all of the circle of
      Willis, placing the arterial wall in the correct location and segmenting an
      acceptable level of noise. 101

5.15 A transverse MIP illustrating (a) a freehand ROI drawn around some
      unwanted data and (b) the data deleted from the MR dataset. 102

5.16 A sagittal MIP illustrating (a) a freehand ROI drawn around the posterior
      of the CoW, (b) the data within the ROI separated from the rest of the
      MR dataset and (c) a coronal view of the separated region of the MR dataset.102

5.17 A transverse slice illustrating (a) a freehand ROI drawn around some un-
      wanted data and (b) the data deleted from the MR dataset. 103

5.18 The Gaussian function in 1D. 104

5.19 A transverse MIP illustrating a freehand ROI around the ACA's, (b) a
      coronal MIP of the separated data with another freehand ROI drawn
      around the ACA's and (c) the smoothed dataset within the two ROI's.
      Note: the smooth level has been exaggerated for effect. 104

5.20 A sagittal MIP illustrating (a) the RPCoA selected with a freehand ROI,
      (b) the RPCoA separated from the rest of the dataset, (c) a transverse
      MIP of the separated RPCoA, (d) the colourmap changed from greyscale
      to RGB, (d) a second freehand ROI drawn around the RPCoA and (d)
      the result after applying a brightening algorithm. 106

5.21 A surface triangulation of the arterial wall of the circle of Willis illustrating
      (a) unwanted noise in the triangulation, (b) a triangular facet selected
      on the carotid siphon (highlighted within blue circle) and (c) the noise
      removed with the neighbour painting filter. 107

5.22 The implementation of the neighbour painting filter algorithm with the
      geometrical entities highlighted by an example. 108

5.23 The right MCA of a circle of Willis model illustrating (a) the hollow artery,
      resulting from the isosurfacing of the main dataset, (b) the interface sur-
      face triangulation created by the isosurfacing of a boundary matrix and
      (c) the porous block and efferent outlets created by the isosurfacing of a
      porous block dataset and copying and translating the triangulation inter-
      face respectively. 109

5.24 (a) A coronal view of the circle of Willis triangulation, (b) the stenosis
     blocks extruded from the afferent terminations of the model, (c) a single
     stenosis block separated and highlighted and (d) the vertices of its surface
     triangulation deformed to simulate a stenosis. 111

5.25 (a) A transverse MIP of the raw MRI data (b) the final segmented model. 112

5.26 Axial views of the (a) missing PCoA and (b) fused ACA patient specific
     circle of Willis anatomical variations created. 113

5.27 An example of the adaptive cartesian mesh created for the segmented
     circle of Willis models. 114

5.28 Tetrahedral, hexahedral, wedge and pyramid cell types used in unstruc-
     tured grids. 115

6.1 The experimental data of Newell et al [112] showing (a) the reduction in
     $MABP$ brought about by a thigh cuff manoeuver and (b) the its effect on
     CBF, illustrating the return to normal by a reduction in CVR. 119

6.2 Autoregulation curve used in the cerebral autoregulation model of Hunts-
     man et al [77]. 121

6.3 A 1D electrical network analogy used to determine the set point flowrates
     through the efferent arteries of the circle of Willis. Assuming a total inflow
     of 750mL/min and a resistance ratio of 6:3:4 between the anterior, middle
     and posterior cerebral arteries respectively. 125

6.4 (a) A generic plot of the sigmoidal function incorporated in cerebrovascular
     resistance limits and (b) the effect on the autoregulation curve. 131

6.5 A sample result from the study of Strandgaard et al, illustrating lower half
     of an autoregulation curve from one volunteer of the hypertensive subgroup
     examined. 133

6.6   (a) A transverse reference MIP showing the placement of a scanning plane
      for determination of the blood flow through the LMCA, (b) the magnitude
      image of the PC scan normal to the scanning plane and (c) the phase image,
      illustrating the LMCA velocity as a white circle. 138

6.7   A transverse scanning plane acquiring velocity data through the internal
      carotid and basilar arteries, showing (a) the phase image and (b) the mag-
      nitude image. 140

6.8   (a) A rectangular ROI drawn around the arteries of interest in a given PC
      dataset and (b) the reduced dataset. 140

6.9   The reduced PC dataset of an example LMCA illustrating (a) the greyscale
      colourmap, (b) and RGB colourmap and c) the interpolated image, giving
      a much clear visualization of the velocity profile. 141

6.10 A surface plot of the LMCA illustrating (a) the coarse waveform and (b)
      the interpolated waveform. The key point of the comparison is that firstly
      the interpolation of the data has negligible effect on the velocity profile
      and b) the region of negative velocities surrounding the lumen of each artery.141

6.11 The reduced PC dataset of an example LMCA illustrating (a) the data
      cleared outside of a rectangular ROI around the lumen with the zero level
      visible as light orange, (b) the zero level becoming a darker orange as the
      negative pixels surrounding the lumen are removed and (c) the isolated
      velocity profile. 142

7.1   The Non Newtonian Effects of Blood [80]   150

7.2   Transient flowrates through the three afferent arteries showing (a) the plot
      of relatively jagged waveforms resulting from the 30 points sampled in the
      cardiac cycle and (b) the plot of the smoothed waveforms. The mean
      flowrates over the cardiac cycle are shown in the top right hand corner of
      each plot. 155

7.3   (a) An example of the pressure waveforms in the three afferent arteries
      and (b) the comparison between the efferent flowrate waveforms calculated
      from MRI to that produced by numerical simulation. 158

7.4   The simulation procedure for performing the circle of Willis CFD simulations.159

8.1   Transient flowrates through the (a) efferent arteries, (b) communicating
      arteries, (c) afferent arteries, and (d) a streamline plot through the com-
      plete circle of Willis at the end of the simulation. 164

8.2   Steady state (a) afferent and (b) communicating flowrates through the
      complete circle of Willis with its arterial variations, both under normal
      cerebral perfusion pressure conditions and after the pressure drop. Note:
      the lighter shade of a bar illustrates the flowrate under normal conditions
      (denoted 'Start') and the darker shade of a bar illustrates the flowrate in
      response to the pressure drop at the end of the simulation (denoted 'End'). 165

8.3   Transient flowrates through the (a) efferent arteries, (b) communicating
      arteries, (c) afferent arteries, and (d) a streamline plot through the fetal
      P1 circle of Willis at the end of the simulation. 167

8.4   Steady state (a) afferent and (b) communicating flowrates through the
      fetal P1 circle of Willis with its arterial variations, both under normal
      cerebral perfusion pressure conditions and after the pressure drop. Note:
      the lighter shade of a bar illustrates the flowrate under normal conditions
      (denoted 'Start') and the darker shade of a bar illustrates the flowrate in
      response to the pressure drop at the end of the simulation (denoted 'End'). 169

8.5   Transient flowrates through the (a) efferent arteries, (b) communicating
      arteries, (c) afferent arteries, and (d) a streamline plot through the missing
      A1 circle of Willis at the end of the simulation. 171

8.6   Steady state (a) afferent and (b) communicating flowrates through the
      missing A1 circle of Willis with its arterial variations, both under normal
      cerebral perfusion pressure conditions and after the pressure drop. Note:
      the lighter shade of a bar illustrates the flowrate under normal conditions
      (denoted 'Start') and the darker shade of a bar illustrates the flowrate in
      response to the pressure drop at the end of the simulation (denoted 'End'). 173

9.1   (a) The maximum percentage difference in communicating artery flowrate
      between successive mesh refinements of the complete circle of Willis model,
      (b) the LMCA velocity profile immediately distal to the ICA-ACA junction
      for various levels of mesh refinement. 179

9.2   An investigation into the pulsatile effects following a 20mmHg drop in
      afferent pressure for the complete circle of Willis model, illustrating for
      the LACA (a) the afferent pressure in the LICA, (b) the transient CBF,
      (c) the transient C_tCO_2, (d) the transient CVR, (e) the transient OEF
      and (f) the transient O_2 delivery. 181

9.3   Transient flowrates through the three communicating arteries of the com-
      plete circle of Willis model, illustrating a comparison between the pulsatile
      and non-pulsatile flowrates. 182

9.4   A comparison between the non pulsatile and cardiac cycle integrated pul-
      satile flowrates in the efferent and communicating arteries, at the end of
      the 20mmHg afferent pressure drop simulation. 183

9.5   A comparison between the Newtonian and non Newtonian viscosity models
      on communicating artery flowrates following a 20mmHg afferent pressure
      drop for the complete circle of Willis model. 185

9.6   A contour plot of the Carreau-Yasuda non Newtonian viscosity throughout
      the complete circle of Willis model. 186

9.7   The end response to an occlusion of the LICA in combination with other
      unilateral stenoses for the complete circle of Willis (a) the efferent CBF,
      (b) the efferent CVR, (c) the efferent C_tCO_2, (d) the efferent OEF, and
      (d) the efferent O_2 delivery. 190

9.8   The end response to an occlusion of the LICA in combination with other
      unilateral stenoses for the complete circle of Willis illustrating the afferent
      and communicating artery flowrates. 191

9.9   The end response to an occlusion of the RICA in combination with other
      unilateral stenoses for the complete circle of Willis (a) the efferent CBF,
      (b) the efferent CVR, (c) the efferent C_tCO_2, (d) the efferent OEF, and
      (d) the efferent O_2 delivery. 194

9.10 The end response to an occlusion of the RICA in combination with other
      unilateral stenoses for the complete circle of Willis illustrating the afferent
      and communicating artery flowrates. 195

9.11 The end response to an occlusion of the RICA in combination with a
      70% LICA stenosis for the complete circle of Willis model, illustrating (a)
      MABP at the arterial wall and (b) a streamline plot colored by the blood
      velocity. 196

9.12 The end response to an occlusion of the BA in combination with other
      unilateral stenoses for the complete circle of Willis (a) the efferent CBF,
      (b) the efferent CVR, (c) the efferent C_tCO_2, (d) the efferent OEF, and
      (d) the efferent O_2 delivery. 198

9.13 The end response to an occlusion of the BA in combination with other
      unilateral stenoses for the complete circle of Willis illustrating the afferent
      and communicating artery flowrates. 199

9.14 The end response to an occlusion of the LICA in combination with other
      unilateral stenoses for the missing PCoA circle of Willis (a) the efferent
      CBF, (b) the efferent CVR, (c) the efferent C_tCO_2, (d) the efferent OEF,
      and (d) the efferent O_2 delivery. 202

9.15 The end response to an occlusion of the LICA in combination with other
      unilateral stenoses for the missing PCoA circle of Willis illustrating the
      afferent and communicating artery flowrates. 203

9.16 The end response to an occlusion of the RICA in combination with other
     unilateral stenoses for the missing PCoA circle of Willis (a) the efferent
     CBF, (b) the efferent CVR, (c) the efferent C_tCO_2, (d) the efferent OEF,
     and (d) the efferent O_2 delivery. 206

9.17 The end response to an occlusion of the RICA in combination with other
     unilateral stenoses for the missing PCoA circle of Willis illustrating the
     afferent and communicating artery flowrates. 207

9.18 The end response to an occlusion of the RICA in combination with a 70%
     LICA stenosis for the missing PCoA circle of Willis model, illustrating (a)
     MABP at the arterial wall and (b) a streamline plot colored by the blood
     velocity. 208

9.19 The end response to an occlusion of the BA in combination with other
     unilateral stenoses for the missing PCoA circle of Willis (a) the efferent
     CBF, (b) the efferent CVR, (c) the efferent C_tCO_2, (d) the efferent OEF,
     and (d) the efferent O_2 delivery. 211

9.20 The end response to an occlusion of the BA in combination with other
     unilateral stenoses for the missing PCoA circle of Willis illustrating the
     afferent and communicating artery flowrates. 212

9.21 The end response to an occlusion of the LICA in combination with other
     unilateral stenoses for the fused ACA circle of Willis (a) the efferent CBF,
     (b) the efferent CVR, (c) the efferent C_tCO_2, (d) the efferent OEF, and
     (d) the efferent O_2 delivery. 215

9.22 The end response to an occlusion of the LICA in combination with other
     unilateral stenoses for the fused ACA circle of Willis illustrating the affer-
     ent and communicating artery flowrates. 216

9.23 The end response to an occlusion of the RICA in combination with other
     unilateral stenoses for the fused ACA circle of Willis (a) the efferent CBF,
     (b) the efferent CVR, (c) the efferent C_tCO_2, (d) the efferent OEF, and
     (d) the efferent O_2 delivery. 218

9.24 The end response to an occlusion of the RICA in combination with other
     unilateral stenoses for the fused ACA circle of Willis illustrating the affer-
     ent and communicating artery flowrates. 219

9.25 The end response to an occlusion of the BA in combination with other
     unilateral stenoses for the fused ACA circle of Willis (a) the efferent CBF,
     (b) the efferent CVR, (c) the efferent C_tCO_2, (d) the efferent OEF, and
     (d) the efferent O_2 delivery. 222

9.26 The end response to an occlusion of the BA in combination with other uni-
     lateral stenoses for the fused ACA circle of Willis illustrating the afferent
     and communicating artery flowrates. 223

9.27 The end response to an occlusion of the BA in combination with a 70%
     RICA stenosis for the fused ACA circle of Willis model, illustrating (a)
     MABP at the arterial wall and (b) a streamline plot colored by the blood
     velocity. 224

10.1 A sample clinical experiment aimed at validating the cerebral autoregula-
     tion model. In the experiment a volunteer is placed inside an MRI scanner
     and given controlled doses of carbon dioxide while the cerebral blood flow
     is measured with phase contrast MRI. 230

10.2 An example of an extension of the 3D geometry whereby the afferent ter-
     mination begins at the aortic arch. 231

10.3 An example of an extension of the 3D geometry whereby the efferent ter-
     minations extended distally to the circle of Willis. 232

A.1 When placed in an external magnetic field, protons align their spin vector
     with the field in either a parallel, low energy configuration or an anti-
     parallel, high energy configuration. 236

A.2 The net magnetic moment in a spin packet is the vector sum of the spin
     vectors of all the individual protons in the parallel and anti-parallel direc-
     tions. Furthermore the net magnetization vector is defined as the vector
     sum of the magnetization vectors of all the spin packets. 238

A.3 Two frames of reference, useful in MRI. The first is the laboratory frame, which is a fixed 3D cartesian coordinate system with the z axis aligned with the main $\mathbf{B}_0$ field. The second frame of reference is the rotating frame, where the x and y axes rotate about the z axis at the Larmor frequency. . 239

A.4 The RF energy pulse contains an oscillating magnetic field (in the laboratory frame), which creates a torque on the longitudinal magnetization vector, tipping it into the xy plane. The flip angle θ through which the longitudinal magnetization is tipped is proportional to the length of the RF pulse. 241

A.5 After being tipped into the xy plane the longitudinal magnetization returns to its equilibrium value. The spin lattice relaxation time T_1 is the time required for the longitudinal magnetization to recover 63% of its equilibrium value. 242

A.6 After being tipped into the xy plane the transverse magnetization rapidly decays due to micromagnetic inhomogeneities within the particular tissue, causing the spins to dephase. In the laboratory frame the x and y transverse magnetization's are decaying sinusoids. In the rotating frame the transverse magnetization follows an exponential decay. 243

A.7 In the laboratory frame the net magnetization vector follows a helical path in 3D space as the protons return to their equilibrium alignment with the main magnetic field. 244

A.8 If the repetition time (the time between subsequent RF pulses) is short enough that the longitudinal magnetization is not allowed to completely return to its equilibrium value, then a saturation of the magnetization can occur, where the amount of magnetization recovered follows a decaying exponential function. 245

A.9 The actual MR signal measured is an echo, created by rephasing the transverse magnetization. One way to create an echo is to apply a 180° pulse at a time $T_E/2$ creating a spin echo at the time T_E. 246

A.10 A T_1 weighted image employs repetition and echo times, maximizing the effects of the T_1 characteristics of the tissue and minimizing the T_2 characteristics [143]. 247

A.11 (a) a T_1 weighted image (b) a T_2 weighted image and (c) a proton density weighted image of a given slice within the brain. 248

A.12 The linear magnetic field gradients are created using pairs of coils. When an electrical current I is passed through the coil a magnetic field is set up in the space surrounding the coil. By combining coils with opposite polarity a linear magnetic field gradient can be created in a given region inside the MRI scanner. 248

A.13 The magnetic field gradient in the z direction means that the resonant (Larmor) frequency will differ for all protons along the z direction. The application of an RF pulse at a given frequency will therefore only cause resonance and 'select' protons in a given slice along the z axis. The thickness of the slice selected depends upon the strength of the gradient and the transmit bandwidth of the RF pulse. 249

A.14 Nine sample voxels in a selected slice. In the phase encode direction the spins within a voxel develop a phase difference during the time that the phase encode gradient is turned on. This phase difference remains however after the phase encode gradient is turned on. In the frequency encode direction the spins within a voxel precess at different frequency due to the application of the frequency encode gradient. As a result, each voxel in the selected slice has a unique combination of precessional frequency and phase. 250

A.15 The pulse and gradient sequence for a typical spin echo. Illustrated are the relative temporal spacings between pulses and applications of the magnetic field gradients. 251

A.16 K space is a complex frequency space where the k space coordinates define spatial waves of various frequency and phase, (a)a k space matrix with the origin at zero frequency and phase (b) a k space image and (c) the resulting MR image. 252

A.17 The MR signal undergoes a process called complex demodulation, using a combination of analogue electronics and computer hardware. The basic steps are separating the raw MR signal, removing the ω_0 carrier frequency, filtering and digitally sampling the signal and finally recombining the two signals to form a complex signal. 253

A.18 An example of the acquisition of a line of k space data. The constant read gradient defines a linear trajectory along k_x as the signal is sampled in time. The sampled signal represents the two signals from the complex demodulation process which are entered into k space as complex numbers. 255

A.19 The MR signal entered into a given k_x line in k space includes the signal from the whole selected slice in both the frequency and phase encode directions. In order to separate the signal in terms of frequency and phase to generate the final MR image phase encode steps of differing phase encode gradient magnitudes are required. Along a given line in k_y, the oscillations in phase form a complex signal (similar to the MR signal read along k_x) allowing the Fourier transform to separate out the different phases. 256

A.20 The time of flight principle. Stationary tissue or blood will experience multiple RF pulses and will become saturated, giving little or no MR signal. Moving blood will not experience same number of RF pulses and will give a stronger signa.l 259

A.21 In order to suppress the signal from venous blood a region above the selected slab has an RF pulse applied to it to saturate the magnetization in the venous blood. As the venous blood enters the excited slab it will hence give little signal. Arterial blood flowing in the opposite direction will not experience the saturation band (before entering the selected slab that is) and will still give a strong signal. 260

A.22 Pulse and gradient sequence for 3D time of flight imaging uses the wide bandwidth RF pulse to excite a thick slab. In this case the k space becomes three dimensional with k_z being the third dimension. The pulse sequence uses the slice select gradient as a phase encode gradient to control the trajectory through k_z. In addition, the read gradient is used to generate the echo in the MR signal. 261

A.23 (a) A time of flight image for a transverse slice, illustrating the enhanced signal from the flowing blood and (b) a maximum intensity projection illustrating the arteries comprising the circle of Willis. 262

A.24 Phase accumulation for both stationary and constant velocity spins during a bipolar gradient. At the center of the echo the phase accumulated by stationary spins is zero, but for spins moving with constant velocity the phase accumulated in non-zero and is proportional to the velocity of the flow. 263

A.25 Pulse and gradient sequence for a phase contrast scan which uses data from two applications of the bipolar gradient (applied in two separate acquisitions), to eliminate phase accumulation due to magnetic field inhomogeneities. 265

A.26 Variations of the three magnetic field gradients may be combined to create slice select planes at any orientation, relative to the laboratory frame. Typically the TOF data will be used to determine the positioning of the plane. Also shown is a schematic representation of the part of the pulse and gradient sequence used to generate the imaging plane. 266

A.27 An example of a phase contrast scan of some blood vessels showing (a) A magnitude and (b) phase image. 267

B.1 The basis for marching cubes illustrated with (a) a stack of MRI images from a 3D TOF scan forming the 3D array M, (b) two consecutive slices in the array with the 4 voxels used per slice highlighted, (c) the eight array values used to form the vertices of the cube with their corresponding assigned values from the 3D array M. 271

B.2 The 15 families used in the original marching cubes algorithm [92]. These families account for every possible way that a cube can be triangulated, if rotational and complementary symmetries are considered for each case. The light blue and dark blue vertices represent vertex values that lie on opposite sides of the isosurface values (i.e. dark blue is above and light blue below, or vice versa). 273

B.3   (a) An example of a general case where the surface triangulation between
      adjacent cubes produces a continuous surface triangulation and (b) an ex-
      ample of an ambiguous case which can occur with the 15 families. Although
      these two cases triangulate their own cube appropriately, the triangulations
      do not connect across the cubes, resulting in a surface triangulation with
      gaps in it.   274

B.4   (a) The labelling scheme used for a standard cube, where E designates an
      edge and V a vertex (b) an example of the left edges L_1, L_2, L_3 and right
      edges R_1, R_2, R_3 for Edge 1. 275

B.5   Two example cube configurations illustrating the surface triangulation and
      the edge intersection distances d_n for each. (a) is the simplest triangulation
      possible with $N_{valid} = 3$ and (b) is one of the more complex triangulations
      with $N_{valid} = 6$.   277

B.6   The procedure for adding edges to the polygon string P sequentially exam-
      ining the next edges of the current edge until another valid edge is found.
      The process repeats until the edge at the beginning of the polygon string
      is found. (a) The polygon string is in fact a single triangle with $N_{edges} = 3$
      and (b) the polygon string is an arbitrary polygon with $N_{edges} = 6$. 279

B.7   The procedure for tessellating a polygon string with more than 3 edges.
      3 vertices v_1, v_2, v_3 and are initialized with P_1, P_2, P_3 and a triangle is
      generated with these vertices. The algorithm then loops as many times as
      there are edges remaining in the polygon string, assigning the value of v_3
      to v_2, reassigning v_3 to the next unused edge in P and adding the resulting
      triangle to the $Faces$ and $Vertices$ arrays. 280

B.8   An outline of the basic steps performed in the marching cubes algorithm
      used in the present study.   281

B.9   An example of a surface triangulation of a 3D TOF array, illustrating a
      portion of the circle of Willis.   282

C.1   Eulerian and Lagrangian viewpoints applied to a control volume and an
        infinitesimal fluid element. The Lagrangian viewpoint involves tracking
        the respective region of fluid from the left to right with the flow and no
        mass crosses the system boundary. The system boundary with the Eulerian
        description is fixed in space and fluid moves into and out of the system. . . 285

C.2   A flow field moving through a control volume. 287

C.3   Stress components acting on an infinitesimal fluid element. 291

C.4   Relative displacement between two points, A and B within a flow field. . . 292

C.5   Types of Motion a Fluid Element May Undergo. (a) Translation, (b)
        Rotation, (c) Dilation and (d) Distortion 295

C.6   Tetrahedral, hexahedral, wedge and pyramid cell types used in unstruc-
        tured grids. 303

C.7   An example portion of a hexahedral grid, illustrating the relationship be-
        tween cell centroids, face centroids, and vertices. 305

C.8   The effect of convective on the propagation of ϕ through the domain. . . . 306

C.9   Two tetrahedral cells (depicted in 2D for simplicity), illustrating the area
        and unit vectors used in the calculation of the diffusive term. 308

C.10 Tetrahedral cells depicted in 2D for simplicity, illustrating the relation be-
        tween the vertex values of c_p and the values at the centroids of its neighbors
        in order to calculate the face value of ϕ. 310

C.11 1D cells within a fluid domain, illustrating the cell centroids, face centroids
        and the face Area vectors with respect to the cell c_0. 316

C.12 When using co-located grids, *Checkerboarding* may result, whereby a highly
        oscillatory and discontinuous pressure field may be obtained. Note: the
        pressure values shown are arbitrary and only used to illustrate the point . 318

C.13 The procedure used in the solution of the governing equations. 325

C.14 The multigrid method solves the governing equations on a series of sequentially coarser grids during the iteration process, relating the values
computed on the coarse grid, back to the fine grid. 326

C.15 Two tetrahedral cells (depicted in 2D for simplicity), in the fine level grid,
whose equations are to be agglomerated into a single equation on the coarse
level grid. 328

List of Tables

1.1   Tabulated values of the anatomic configurations in the anterior circle of Willis found in the study by Hartkamp et al [58]. 15

1.2   Tabulated values of the anatomic configurations in the posterior circle of Willis found in the study by Hartkamp et al [58]. 15

1.3   Anatomic configurations of the circle of Willis found in the study by Alpers et al [12]. 17

1.4   Anatomic configurations of the circle of Willis and associated neuro-logic dysfunction found in the study by Alpers et al [11]. 18

5.1   Circle of Willis measurements used for the creation of the idealized geometries. Note: *ips* and *cont* are abbreviations for ipsilateral and contralateral. 93

5.2   Time of Flight parameters used in the acquisition of the MRI scans. Note: the definition of these parameters is given in Appendix A. 95

6.1   Constant parameters used in the oxygen and carbon dioxide concentration model of Dash et al [35] and incorporated in the present study. 129

6.2   CBF and diameter values in the major cerebral arteries measured by Enz-mann et al [43]. 137

6.3   Phase Contrast parameters used in the acquisition of the MRI scans. Note: the definition of these parameters is given in Appendix A. 139

9.1   The efferent flowrates for the complete circle of Willis measured with PC
      MRI, and the values of CVR required to achieve them in the numerical
      simulation. 187

9.2   The efferent flowrates for the missing PCoA circle of Willis measured with
      PC MRI, and the values of CVR required to achieve them in the numerical
      simulation. 200

9.3   The efferent flowrates for the fused ACA circle of Willis measured with PC
      MRI, and the values of CVR required to achieve them in the numerical
      simulation. 213

A.1   Relaxation times for important biological tissues [143]. 245

B.1   The edge lookup table used in the present algorithm. Each edge is defined
      by its two vertices and its left and right edges. 276

B.2   The coordinate lookup table used in the present algorithm to convert the
      intersection distance along an edge to 3D coordinates based on the i, j, k
      position of the cube within the 3D array. 280

Nomenclature

Mathematical Symbols

a_c	Centre coefficient of the property ϕ_c for a control volume c
a_n	Coefficient of the property ϕ_c for a neighbouring control volume n
A	Linearized coefficient matrix for all control volumes within a computational grid
A_k	Fourier coefficient
$\mathbf{A}$	Area vector
$\mathbf{A}_f$	Area vector of a face f within a computational grid
ABP	Arterial blood pressure
Aut_{CVR}	Autoregulation activation for CVR
Aut_{OEF}	Autoregulation activation for OEF
b_c	Known terms of the property ϕ_c a control volume c
$\mathbf{B}_0$	Main magnetic field
c	A cell within a computational grid
C	An array of cells in the context of the finite volume method or the constant in the CVR dynamic for the patient specific autoregulation model
C_aCO_2	concentration of carbon dioxide in arterial blood
C_aO_2	concentration of oxygen in arterial blood
CBF	Cerebral blood flow
CBF_{SP}	Set point for cerebral blood flow
CBV	Cerebral blood volume
$CMRO_2$	Cerebral metabolic rate for oxygen consumption
CO	Cardiac output

$C_t CO_2$	concentration of carbon dioxide in the tissue
$C_v O_2$	concentration of oxygen in venous blood
CVR	Cerebrovascular resistance
CVR_{SP}	Set point for cerebrovascular resistance
D_f	Primary diffusion term at a face f within a computational grid
E_f	Secondary diffusion term at a face f within a computational grid
f	A face within a computational grid
F	An array of faces within a computational grid or surface triangulation
G	Gaussian kernel
G_x	Magnetic field gradient in the x direction
G_y	Magnetic field gradient in the y direction
G_z	Magnetic field gradient in the z direction
h	Plank's constant
HR	Heart rate
i	Array counter in the x direction
I	Identity matrix
ICP	Intracranial Pressure
j	$\sqrt{-1}$
k	Boltzmann's constant
k_x	x direction in k space
k_y	y direction in k space
k_z	z direction in k space
K	Total number of Fourier coefficients
l	Multigrid level
m	Iteration level
M	3D array resulting from a time of flight MRI scan
M_x	x component of the net magnetization vector
M_y	y component of the net magnetization vector
M_z	z component of the net magnetization vector
M_{z0}	equilibrium net magnetization vector
$\mathbf{M}$	Net magnetization vector
$MABP$	Mean arterial blood pressure
n	Neighbouring control volume to a control volume c within a computational grid

O	Constant in the OEF dynamic for the patient specific autoregulation model
OEF	Oxygen extraction fraction
p	Thermodynamic pressure
p^*	Initial pressure
p'	Pressure correction
P	A collection of edges forming a closed polygon
P_aCO_2	Partial pressure of carbon dioxide in arterial blood
P_aO_2	Partial pressure of oxygen in arterial blood
PR	Peripheral resistance
P_tO_2	Partial pressure of oxygen in the tissue
q_c	Centre coefficient of the pressure correction p'_c for a control volume c
q_n	Coefficient of the pressure correction p'_n for a neighbouring control volume n
Q	Volume flowrate
ROI	Region of interest
S_aCO_2	Carbon dioxide saturation of hemoglobin in arterial blood
S_aO_2	Oxygen saturation of hemoglobin in arterial blood
SV	Stroke volume
T	Temperature
T_E	Echo time
T_R	Repetition time
T_1	Spin lattice relaxation time
T_2	Spin-spin relaxation time
T_2^*	Spin-spin relaxation time including the effects of magnetic field inhomogeneity
u	x component of the velocity vector $\mathbf{u}$
$\mathbf{u}$	3D velocity vector
v	y component of the velocity vector $\mathbf{u}$
V	An array of vertices within a computational grid or surface triangulation
V_c	the volume of the control volume c within a computational grid
w	z component of the velocity vector $\mathbf{u}$
α	Under relaxation factor

$\bar{\epsilon}$	Rate of strain tensor
γ	Gyromagnetic ratio
Δt	Time step size
η	Apparent viscosity
θ	Flip angle
λ	Bulk viscosity of a fluid
$\bar{\tau}$	Shear stress tensor
ϕ	Represents either $\mathbf{u}$, or T in the generalized transport equation or phase angle the context of magnetic resonance imaging
Φ	Represents extensive properties of mass, momentum or internal energy
ω	Larmor frequency
$\bar{\omega}$	Rate of rotation tensor

Arterial Acronyms

ACA	Anterior cerebral artery
AChA	Anterior choroidal artery
AICbA	Anterior inferior cerebellar artery
AIFA	Anterior frontal artery
AParA	Anterior Parietal artery
ASpA	Anterior spinal artery
ATA	Anterior temporal artery
AngA	Angular artery
BA	Basilar artery
CCA	Common carotid artery
CMA	Callosomarginal artery
CalA	Calcarine artery
LA	Labyrinthine artery
LFbA	Lateral frontobasal artery
LenStrA	Lenticulostriate arteries
MCA	Middle cerebral artery
MFbA	Medial frontobasal artery
MIFA	Middle frontal artery
MStA	Medial striate artery
MTA	Middle temporal artery

OPhA	Ophthalmic artery
PCA	Posterior cerebral artery
PCeA	Paracentral artery
PCnA	Precuneal artery
PICbA	Posterior inferior cerebellar artery
PIFA	Posterior frontal artery
PLChA	Posterior lateral choroidal artery
PMChA	Posterior medial choroidal artery
PParA	Posterior parietal artery
PTA	Posterior temporal artery
ParOccA	Parietooccipital artery
PerA	Pericallosal artery
PonA	Pontine arteries
PreCenA	Precentral sulcal artery
PreFrA	Prefrontal artery
SCbA	Superior cerebellar artery
ToccA	Temporooccipital artery
TpolA	Temporopolar artery
VA	Vertebral artery
CenA	Central sulcal artery
ECA	External carotid artery
FpA	Frontopolar artery
ICA	Internal carotid artery

Abstract

The Circle of Willis (CoW) is a ring-like arterial structure forming the major anastomotic connection between arterial supply systems in the brain, and is responsible
for the distribution of oxygenated blood throughout the cerebral mass. Among the
general population, only approximately 50% have a complete CoW, where absent
or hypoplastic vessels are common among a multitude of possible anatomical variations, reducing the degree to which blood may be rerouted. While an individual with
one of these variations may under normal circumstances suffer no ill effects, there
are certain pathological conditions which can present a risk to the person's health
and increase the possibility of suffering an ischaemic stroke when compounded with
an anatomical variation. This body of work presents techniques for generating 3D
models of the cerebral vasculature using magnetic resonance imaging (MRI) and
performing computational fluid dynamics (CFD) simulations in order to simulate
the flow patterns throughout a circle of Willis. Incorporated with the simulations
is a mathematical model of the cerebral autoregulation mechanism, simulating the
ability of the smaller arteries and arterioles in the brain to either constrict or dilate
in response to alterations in cerebral blood flow, thereby altering the cerebrovascular resistance of each major brain territory and regulating the amount of blood
flow within a physiological range of cerebral perfusion pressure. The CFD simulations have the ability to predict the amount of collateral flow rerouted via the
communicating arteries in response to a stenosis or occlusion, and the major ob-
jective of this study has been the investigation of how anatomical variations of the
circle of Willis affect the capacity to provide this collateral flow. Initial work began
with the development of three idealized models of common anatomical variations,
created using computer aided design software (CAD) and based on the results of
MRI scans. The research then shifted to developing a technique whereby patient
specific models of the circle of Willis could be directly segmented from the MRI
data. As a result of this shift, an interactive GUI-based tool was developed for the
processing of the MRI datasets, allowing for rapid data enhancement and creation

of a surface topology representing the arterial wall of the circle of Willis, suitable for a CFD simulation. The results of both sets of simulations illustrate that there exist a number of variables associated with a patients circle of Willis geometry, such as cerebral blood flow and combinations and degrees of stenosis, implying that the initial goal of drawing generalized conclusions was perhaps flawed. Instead, a crucial outcome of this body of work is that the future research should be directed toward extending the physiological complexity of both the geometry and the autoregulation model, with the intention of a patient specific application rather than producing large datasets with which to make broad generalizations.

Chapter 1

Anatomy

1.1 Introduction

The application of computational fluid dynamics (CFD) to the modelling of cerebral
hemodynamics requires some background into the anatomy, physiology and pathology of the cerebral vasculature, before any modelling decisions are to be made. The
purpose of the first three chapters will be to provide this information, maintaining
these three distinctions relevant to cerebral hemodynamics. This chapter will begin
by introducing the arteries that deliver blood into the brain and the arteries which
distribute this blood throughout the cerebral territories, subsequently introducing
the structure of the circle of Willis. The presence of *anastomotic connections* (com-
munication between blood vessels by means of collateral channels) between certain
vascular beds will be addressed, followed by a review of studies investigating the
anatomical variability of the circle of Willis. Finally a basic outline of blood vessel structure will be given, as it will integrate with a number of the physiological
processes outlined in Chapter 2.

Before proceeding, some anatomical terminology must be defined to aid in the
description of the cerebral vasculature, which will also be adopted throughout this
text. The terms left and right are used in their normal sense describing a structure
on the individuals left or right side. The terms *inferior* and *superior* are synonymous
with 'lower' and 'higher' respectively and would be used to describe structures when
the individual is standing upright in the anatomical position (Figure 1.1). The terms
anterior and *posterior* are synonymous with 'before' and 'following' respectively
and when in the anatomical context refer to the 'front' and 'back' of a person. A
transverse plane is parallel to the ground and divides the body into inferior and

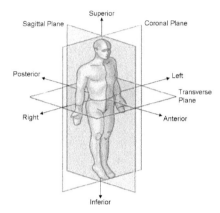

Figure 1.1: A description of terminology useful in describing structures within the body, while an individual is in the anatomical position.

superior structures. A *sagittal* plane runs vertically through the body, dividing it into left and right sides. If the sagittal plane runs through the midline then the plane is a *midsagittal* plane. A *coronal* plane is normal to both the transverse and sagittal planes and vertically through the body, dividing it into anterior and posterior. The terms *medial* and *lateral* refer to a location towards or away from a midsagittal plane. The terms *proximal* and *distal* are used to describe a structure which is closer or farther to the point of attachment to the body with reference to another structure (e.g. the hand is distal to the elbow). The terms *afferent* and *efferent* mean directed toward and away from an organ or structure respectively. When used in the context of describing cerebral arteries, an afferent artery carries blood into the cerebral territory, whereas an efferent artery carries blood out to the distal regions of the brain. The terms *ipsilateral* and *contralateral* refer to structures on the same side and opposite side respectively.

1.2   Cerebral Arteries

It is important to realize that the cerebral vasculature is a highly complex and variable structure and while the larger arteries between the heart and the proximal brain tend to follow similar patterns between individuals, the further into the cardiovascular system one delves, the more variable and difficult to classify the arteries become. This outline will therefore consider the more important cerebral arteries, providing a generalized structure compiled from a number of anatomical studies [51, 85, 124, 140]. Distal to the terminations of the arteries considered, are arteriole and capillary beds perfusing a particular structure in the brain. When a terminal artery is reached in this discussion, the region of the brain that it perfuses with blood will be given, where a detailed introduction to the regions of the brain can be found in [149].

Immediately distal to the heart is the *aortic arch*, from which the *Brachio-cephalic*, *Left Common Carotid* and *Left Subclavian* arteries bifurcate respectively. Immediately distal to the start of the Brachiocephalic artery, it bifurcates into the *Right Common Carotid* and *Right Subclavian Arteries*. Afferent blood supply into the brain occurs through the these left and right common carotid arteries (CCA's) as well as through the left and right Vertebral Arteries (VA's) which bifurcate off their respective subclavian artery (Figure 1.9 and 1.11). All four arteries course superiorly and near the superior end of the neck, at a level known as the *carotid bifurcation*, the common carotid arteries bifurcate into the *External Carotid Artery (ECA)*, which courses into the facial tissue to perfuse it with blood, and the *Internal Carotid Artery (ICA)*, which courses into the brain to supply it with blood.

Considering first the path of the ICA, the first artery branching from it following the carotid bifurcation is the *ophthalmic artery (OPhA)* which supplies blood to the eye and structures of the face. The second artery is the *anterior choroidal artery (AChA)* which supplies blood to the *optic tract, lateral geniculate body, amygdala, hippocampus, internal capsule, lenticular nucleus* and the *thalamus*. The ICA then bifurcates into the *anterior cerebral artery(ACA)* and the *middle cerebral artery (MCA)* (Figure 1.2).

Following the path of the ACA, it courses anteriorly at the bottom of the *longi-tudinal fissure*. The first efferent artery is the *medial striate artery (MStA)* (Figure 1.2) which supplies blood to the *caudate nucleus, internal capsule* and the *putamen*. The next efferent artery is the *medial frontobasal artery (MFbA)* (Figure 1.3) which supplies blood to the anterior *frontal lobe*. Distal to the MFbA the ACA courses superiorly within the *longitudinal fissure* and gives rise to the *Frontopolar Artery (FpA)*, supplying blood to the anterior frontal lobe (Figure 1.3). The ACA courses posteriorly around the *genu* of the *corpus callosum* and then bifurcates into the

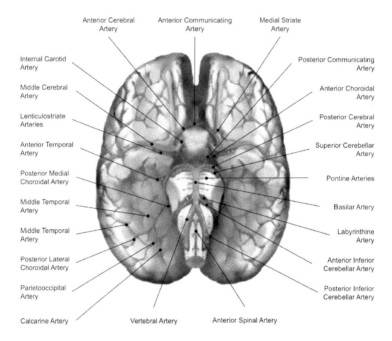

Figure 1.2: An inferior transverse view of the brain, illustrating the afferent and efferent arteries visible from this viewpoint.

callosomarginal artery (CMA) and the *pericallosal artery (PerA)* (Figure 1.3). The CMA gives rise to four arteries known as the *anterior, middle* and *posterior Internal Frontal Arteries (AIFA, MIFA* and *PIFA* respectively), as well as the *Paracentral Artery (PCeA)*, all of which supply blood to the medial *cortex* of the frontal lobe (Figure 1.3). The PerA is essentially a continuation of the ACA, which continues posteriorly with the corpus callosum and giving rise to the *Precuneal Artery (PCnA)* (Figure 1.3), supplying the medial cortex of the frontal and *parietal* lobes. The terminal branches of the PerA anastomose with the distal branches of the middle and posterior cerebral arteries.

The MCA is essentially a continuation of the ICA, perfusing the greatest proportion of the brain with blood compared to the ACA and PCA. The MCA courses laterally towards the surface of the cerebral cortex, the first branches originating from it being the *Lenticulostriate Arteries (LenStrA)* (Figure 1.2), usually 10 to 20

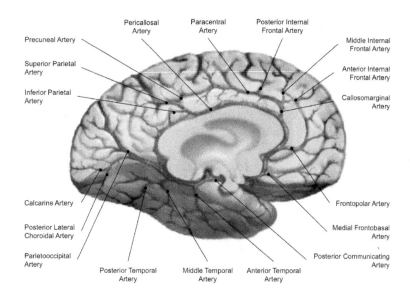

Figure 1.3: A midsagittal view of the brain, illustrating the afferent and efferent arteries visible from this viewpoint.

in number, which supply blood to the *putamen, globus pallidus, head of the caudate nucleus* and the *internal capsule*. As the MCA reaches the lateral fissure it bifurcates into two major branches, known as the *inferior* and *superior* trunks, which course inferiorly and superiorly over the surface of the *cerebrum* respectively. Following the path of the inferior trunk, the first divisions are the *Temporopolar Artery (TpolA)*, the *Anterior, Middle* and *Posterior Temporal Arteries (ATA, MTA* and *PTA* respectively), which course inferiorly over the surface of the cerebral cortex and supply blood to the temporal lobe (Figure 1.4). Distal to these branches are the *Temporooccipital Artery (TOccA)*, the *Angular Artery (AngA)* and the *Posterior Parietal Artery (PParA)*, which supply blood to the posterior *parietal lobe* (Figure 1.4). Following the path of the superior trunk, the first division is the *Lateral Frontobasal Artery* (LFbA), which courses anteriorly on the inferior surface of the cerebral cortex and supplies blood to the middle and inferior *frontal gyrus* and the *frontal basal area* (Figure 1.4). The next division is the *Prefrontal Artery (PreFrA)*, which courses anteriorly and superiorly on the surface of the cerebral cortex, supplying blood to the frontal lobe. Distal to these divisions are the *Precentral* and

Central Sulcal Arteries (*PreCenA* and *CenA* respectively), which course superiorly over the surface of the cerebral cortex. The PreCenA supplies blood to *Broca's area* in the inferior *frontal gyrus* and the *precentral gyrus* and the CenA supplies blood to the *postcentral gyrus* and the superior parietal lobe. The final division which will be covered is the *Anterior Parietal Artery (AParA)* which courses superiorly over the surface of the cerebral cortex and supplies blood to the middle parietal lobe (Figure 1.4). The terminal branches of the MCA, coursing over the cerebral cortex proceed to anastomose with the terminal branches of both the anterior and posterior cerebral arteries.

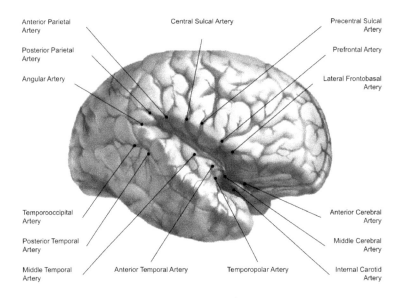

Figure 1.4: A sagittal view of the brain, illustrating the afferent and efferent arteries visible from this viewpoint.

Considering now the course of the VA, the first major artery to branch off as it courses superiorly on the anterior surface of the spinal cord is the *Posterior Inferior Cerebellar Artery (PICbA)* (Figure 1.2). The PICbA supplies blood to the inferior surface of the *cerebellar hemispheres* and the lateral *medulla oblongata*. The next artery branching off the VA is the *Anterior Spinal Artery (ASpA)* (Figure 1.2). The two ASpA's, one arising from each VA, then fuse together to form a single ASpA, which then courses inferiorly on the anterior surface of the spinal cord, supplying

blood to the majority of the spinal cord. As the two VA'S reach the junction
between the *medulla oblongata* and the *pons* they fuse to form a single *Basilar
Artery (BA)* which courses superiorly on the anterior surface of the pons. The first
arteries arising from the BA are the *Anterior Inferior Cerebellar Arteries (AICbA)*
(Figure 1.2), which supply blood to the anterior inferior *cerebellum* and the inferior
pons. The next arteries arising from the BA are the *Labyrinthine Arteries (LA)*
which supply blood to the middle ear. Subsequently, a number of small arteries
branch off the BA, collectively known as the *pontine arteries (PonA)* which supply
the pons. The next arteries given off are the *Superior Cerebellar Arteries (SCbA)*,
which bifurcate into the *Lateral Meningeal* and *Superior Vermian Arteries* which
supply blood to the cerebellar cortex, the inferior *midbrain* and the anterior pons.
Immediately after the SCbA's at the midbrain, the BA bifurcates into two *posterior
cerebral arteries (PCA's)* (Figure 1.2) which curve around the midbrain coursing
posteriorly.

The PCA segment courses laterally around the midbrain, and from each stems
the *Posterior Medial Choroidal Artery (PMChA)* (Figure 1.2) which supplies blood
to the *third* and *fourth ventricle*. The next divisions from the PCA are the *Anterior,
Middle* and *Posterior Temporal Arteries (ATA, MTA* and *PTA* respectively) which
supply blood to the interior surface of the anterior, middle and posterior temporal
lobe. It should be noted that while these names have been applied to the arteries
branching off from the inferior trunk of the MCA, the distinction is made between
them based upon their artery of origin. Distal to these arteries is the *Posterior
Lateral Choroidal Artery (PLChA)* which supplies blood to the posterior *thalamus*
and the lateral *ventricular choroid plexus*. The final branches to be considered in
this discussion are the *Parietooccipital Artery (ParOccA)* and the *Calcarine Artery
(CalA)* which supply blood to the *occipital* lobe (Figure 1.2).

To aid in the reinforcement of the various arteries of the cerebral vasculature just
outlined, Figure 1.5 provides a schematic of the complex 3D anatomical structure
projected onto a 2D plane. Also shown are the general cerebral territories supplied
by each artery.

At the level of the midbrain, just below the hypothalamus is the major anas-
tomosis in brain, known as the *circle of Willis (CoW)* (Figure 1.6). The circle of
Willis is an arterial ring and is completed by the *Anterior Communicating Artery
(ACoA)* and two *Posterior Communicating Arteries (PCoA's)* . The ACoA con-

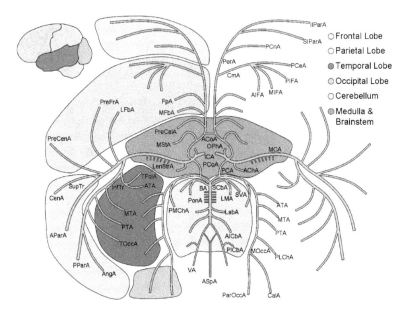

Figure 1.5: A schematic of the cerebral arteries, labelled by their abbreviations and illustrating the general regions of the brain which they supply with blood.

nects the two ACA's where the segment of the ACA, proximal to the ACoA is known as the *A1* segment and the segment distal to the ACoA is known as the *A2* segment. The PCoA connects the ICA to the PCA on the left and right sides of the brain respectively, where the segment of the PCA proximal to the PCoA is known as the *P1* segment and the segment distal to the PCoA is known as the *P2* segment. The numbering scheme for portions of a cerebral artery can be extended further downstream of the circle of Willis to include all of the arteries outlined previously as belonging to a certain segment of its cerebral artery of origin (e.g. the A3, A4, A5 segments of the ACA). While these definitions need not be given here it should be noted that the portion of the MCA immediately distal to the CoW is known as the *M1* segment of the MCA.

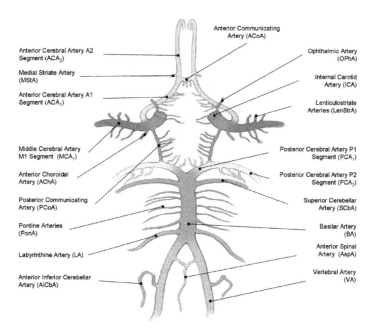

Figure 1.6: The cerebral arteries comprising and immediately surrounding the circle of Willis.

1.3 Collateral Circulation

Collateral pathways for blood flow exist as anastomotic connections between supply systems and provide a means for blood to be rerouted if the primary blood supply to a given region of the brain is reduced. As will be discussed in Chapter 3, they can become important in situations when a larger afferent artery, or even one of the smaller cerebral arteries, becomes blocked. The effectiveness of collateral circulation depends upon a number of factors including the size of the anastomotic vessels, the length of time involved in the development of an occlusion and the size and location of the stenotic or occluded vessel. The collateral pathways may be classified into four categories [170] as collateral circulation via *parallel pathways*, *transverse anastomoses*, micro networks of *ring anastomoses*, and *capillary anastomoses*. These collateral pathways will now be discussed with reference to the occlusion of the afferent arteries, important in the cerebral circulation, considering first the collateral circulation via parallel pathways.

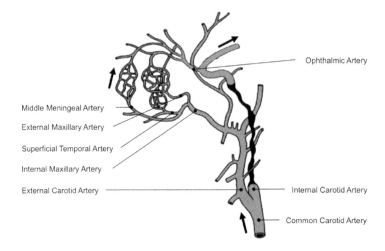

Figure 1.7: Collateral parallel pathway for blood flow given an occlusion of the ICA.

Midway up the neck, the CCA bifurcates into the ICA and the ECA. While the ICA is the major afferent artery in the brain, the external carotid artery supplies blood to the facial structures. Given an occlusion of an ICA however (Figure 1.7) a collateral pathway exists between the ECA and ICA. By increasing blood supply through the ECA, blood may be rerouted through the *Internal Maxillary Artery* and then through the *Middle Meningeal, Superficial Temporal* and *External Maxillary Arteries* to anastomose with the ophthalmic artery, thereby supplying blood to the regions of the brain normally supplied through the ICA.

Given an occlusion of the BA before the SCbA's (Figure 1.8), a collateral pathway exists between AICbA's and the SCbA's. If there is also a bilateral occlusion of the VA's proximal to the origin of the ASpA, an anastomotic branch between the VA and the ASpA will bypass the VA occlusion, supplying blood to the BA for distribution.

Collateral circulation via transverse anastomoses is so named because it constitutes blood flow which is rerouted from anterior to posterior (or vice versa), or from left to right (or vice versa). Given again the occlusion of an ICA (Figure 1.9) there

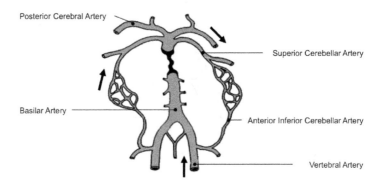

Figure 1.8: Collateral parallel pathway for blood flow given an occlusion of the BA.

exists a collateral pathway through the CoW. By increasing blood supply through the VA's and the contralateral ICA, blood may be rerouted through the PCA's and PCoA's as well as through the ACA's and ACoA.

Given an occlusion of a major cerebral artery, for example the MCA, (Figure 1.10) there exist anastomotic connections between the terminal branches of the ACA's, MCA's and PCA's over the surface of the cerebral cortex. The anastomoses between the MCA and PCA are known as the *meningeal* anastomosis and as the occlusion of the MCA is more common, the direction of blood flow is usually from PCA to MCA. There is also an anastomosis between the two ACA's which occurs along the *corpus callosum.*

Collateral circulation will in general be a combination of both parallel and transverse anastomoses and one such important case is the occlusion of the CCA (Figure 1.11). In this case, blood supply may be increased through the contralateral CCA and subsequently through the *homologous branches* of the ECA. In addition, blood supply may be increased through the ipsilateral subclavian artery and subsequently through the *Deep Cervical, Inferior Thyroid* and *Ascending Cervical Arteries* to anastomose with the ipsilateral ECA, restoring blood supply to the ipsilateral ICA. Furthermore, another pathway exists through the ipsilateral VA, where blood supply may be increased and rerouted through the *muscular branch* of the VA, to anastomose with the ipsilateral ECA.

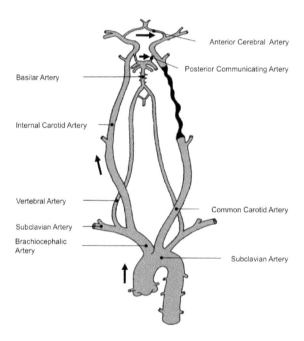

Figure 1.9: Collateral transverse anastomoses for blood flow via the circle of Willis, given an occlusion of the ICA.

The micro network of ring anastomoses exist as connections of arteries on the surface of the brain but do not appear to be very effective in rerouting blood in cases of acute occlusions [170]. Similarly the capillary anastomoses do not seem to contribute any significant blood supply when an intracerebral occlusion occurs. The important point is that there exist a number of different pathways for blood to be rerouted, but by far the greatest source of collateral flow is the circle of Willis, due to the generally larger vessels and shorter distances which blood must travel in order to be rerouted compared to other collateral pathways. Furthermore, it is thought that the ability to use the other collaterals depends on the perfusion pressure at that level, which is determined proximal to these anastomoses at the level of the circle of Willis. However, the anatomy of the CoW differs substantially among the general population in terms of the size and even presence of certain arteries. In many cases these variations can have a detrimental effect on the ability of the CoW to provide collateral flow.

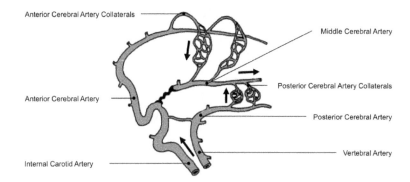

Figure 1.10: Collateral transverse anastomoses between the major cerebral arteries, given an occlusion of the MCA.

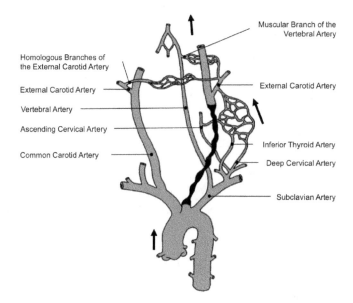

Figure 1.11: Collateral parallel and transverse anastomoses for blood flow, given an occlusion of the CCA.

1.4   Variability of the circle of Willis

There have been a number of studies performed on the anatomy of the circle of Willis, investigating the particular anatomical configuration, dimensions of the blood vessels, and the relationship between circle of Willis geometries and neurological dysfunction among the general population. This discussion will focus on three such studies, two performed on cadaveric brains during autopsy and the other on living individuals using *Magnetic Resonance Imaging (MRI)* (Note: see Appendix A for a detailed outline of MRI).

In the study performed by Hartkamp et al [58], MRI was used to determine the configurations and dimensions of the vessels of the circle of Willis. The study consisted of two groups, designated as control subjects and patients. The patients consisted of seventy five individuals having symptomatic ICA *stenosis* (a partial blockage) and were grouped into four categories; significant unilateral or bilateral stenosis of the ICA, unilateral occlusion of an ICA, unilateral occlusion of an ICA in combination with a significant stenosis of the other ICA, and occlusion of both ICA's. The control subjects consisted of one hundred individuals with various chronic diseases, but no known symptomatic ICA stenosis. As it is only the particular anatomical configurations of the circle of Willis which are of interest for the present study, their results for the dimensions of the vessels will be omitted. The variations were grouped into anterior and posterior variations (Figures 1.12 and 1.13) with the results presented in Tables 1.1 and 1.2.

Considering first the anterior variations, it was found that the patients showed a significantly higher percentage of complete CoW configurations. There was a significantly higher prevalence of two or more ACoA's among the patients and a significantly higher prevalence of a hypoplastic (incomplete or arrested development) or absent ACoA or ACA_1 among the control subjects. The posterior variations showed that again the patients had a significantly higher percentage of complete CoW configurations compared to the control subjects. Both groups showed a significant prevalence of a unilateral hypoplastic or absent PCoA, and slightly less commonly a bilateral hypoplastic or absent PCoA.

The important conclusions drawn from this study were that patients suffering from ICA stenosis showed a higher prevalence of a complete CoW, the implication being that they had only survived the significant level of stenosis (or occlusion) due to the ability of their CoW to provide collateral flow. Furthermore, this study illustrates the prevalence of the anatomical variations of the CoW, with only 55% of the patients having an entirely complete CoW, and only 36% of the control subjects.

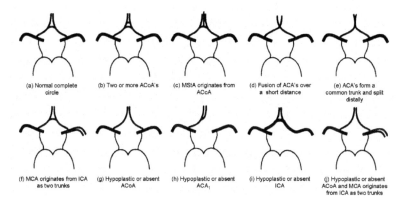

Figure 1.12: Anatomic configurations in the anterior circle of Willis found in the study by Hartkamp et al [58].

Table 1.1: Tabulated values of the anatomic configurations in the anterior circle of Willis found in the study by Hartkamp et al [58].

Group	Number of Cases	Complete					Incomplete					Total Complete
		a	b	c	d	e	f	g	h	i	j	
Patients	75	56	21	3	4	0	4	7	5	0	0	88
Controls	100	57	5	0	2	2	2	20	10	0	2	68

Table 1.2: Tabulated values of the anatomic configurations in the posterior circle of Willis found in the study by Hartkamp et al [58].

Group	Number of Cases	Complete			Incomplete							Total Complete
		k	l	m	n	o	p	q	r	s	t	
Patients	75	54	8	1	27	7	1	0	1	0	1	63
Controls	100	25	14	8	30	12	2	5	1	1	2	47

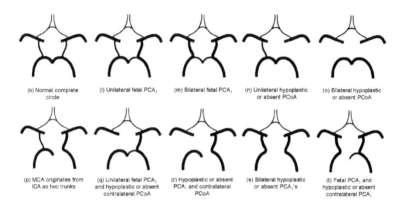

Figure 1.13: Anatomic configurations in the posterior circle of Willis found in the study by Hartkamp et al [58].

In the study performed by Alpers et al [12], eight hundred and thirty seven brains were examined, of which three hundred and fifty were tabulated in terms of anatomical configurations. The findings of the study (Table 1.3) included the presence of variations as either unilateral and bilateral (where appropriate) and recorded the prevalence of a variation in combination with others. The important results of the study are that only approximately 50% of the brains examined possessed a normal complete CoW. Similar to [58], among the posterior variations the most common is a unilateral hypoplastic PCoA with a bilateral hypoplastic PCoA being slightly less common and again this was the most common variation to be found in combination with other variations of the CoW. However, the actual absence of a PCoA was among the least common found. Also similar to [58], a unilateral *fetal P1* (a hypoplastic P1 segment of the PCA) was common among the posterior variations with a bilateral fetal P1 slightly less common and these were the second most common variation found in combination with others. Among the anterior variations, two or more ACoA's was the most common, similar to [58]. The second most common variations were hypoplastic ACoA's, then A1 segments of the ACA, and in concordance with [58], the hypoplastic ACoA was the third most prevalent variation found in combination with others.

In another study performed by Alpers et al [11], seven hundred and sixty two cadaveric brains were examined by autopsy and the relationship between neurologic

Table 1.3: Anatomic configurations of the circle of Willis found in the study by Alpers et al [12].

Variation	Prevalence of Variation %			
	Single		Combined	Total
	Unilateral	Bilateral		
Normal Complete	-	-	-	52.3
Absent left PCoA	0.6	0	0	0.6
Fusion of the ACA's over a short distance	1.4	-	0.3	1.7
Hypoplastic PCoA	14	6	8.3	22.3
Hypoplastic ACoA	0.9	-	2	2.9
Hypoplastic A1 segment of the ACA	0.6	0	1.7	2.3
Two or more ACoA's	2	-	6.6	8.6
Three ACA's	4	0	4	8
Duplicate A1 segment of the ACA	0.9	0	0	0.9
MCA originates from ICA as two separate trunks	0.6	0	0	0.6
Forked PCoA	0.6	0	0	0.6
Duplicate PCA	0.3	0	0	0.3
Hypoplastic P1 segment of the PCA	4.6	3.7	6.3	14.6
Multiple variations	-	-	-	13.4

dysfunction and the configuration of the CoW tabulated. The study consisted of one hundred and ninety four cases of *encephalomalacia* (localized softening of the brain tissue), fifty six of which were *thrombosis*, fifty three *embolism*, and eighty five unknown, one hundred and twenty seven cases of cerebral *hemorrhage*, ninety one cases of cerebral *aneurysm*, and three hundred and fifty cases where the individuals death was not due to neurologic dysfunction, termed 'normal'. Table 1.4 shows the results of their study. The classification of particular CoW variations is not as detailed as the previous two studies, but the results show that similar to the previous two studies there is a significant prevalence of CoW's among the general population possessing some anatomical variation. Furthermore, among the four groups there is a high prevalence of hypoplastic PCoA's and accessory vessels (of which two or more ACoA's is the major contributor). The results also show that the actual absence of a PCoA is quite uncommon, with the hypoplasia of the vessel being much more prevalent. In contrast however, there is a higher occurrence of fetal P1's compared to [58]. In comparison between the variations of the circle of Willis and the neurological dysfunction, the results show that in terms of ischaemic stroke, there were a lower percentage of normal complete CoW's and hence a higher prevalence of absent or hypoplastic vessels compared to the normal brains which showed no neurological dysfunction, the implication being that the deficient CoW's may not have been able to provide adequate collateral flow to protect the individuals studied from ischaemia.

Table 1.4: Anatomic configurations of the circle of Willis and associated neurologic dysfunction found in the study by Alpers et al [11].

| | | Prevalence of Variation % | | | |
Group	Number of Cases	Normal Complete	Hypoplastic PCoA	Absent PCoA	Fetal PCA$_1$	Accessory Vessels
Normal Brains	350	52	22	0.6	15	17
Encephalomalacia	194	33	38	1.5	29	19
Hemorrhage	127	66	17	2.0	5	6
Aneurysm	91	37	40	3.0	22	18

One important point worthy of mention is that in the study of Hartkamp et al [58] the MRI techniques used to observe the cerebral vasculature rely on the movement of flowing blood in order to generate an image. The brightness in an image is degraded by factors such as slow blood flow, when the direction of flow is not normal to the imaging plane, and recirculating regions. Furthermore, the resolution of the MRI images may in fact be lower than the diameters of the hypoplastic vessels in which case they could not be resolved properly. If any of these factors are present, then they may result in the signal from a particular artery being too weak to appear in the angiogram and for that reason the anatomical variations in [58] are grouped into hypoplastic *or* absent, because a given artery may still in fact be present but not appear in the angiogram. Considering the PCoA's, they are generally much smaller compared to the other major cerebral arteries (implying less blood flow through them as well) and are not normal to the imaging plane. For that reason they are harder to observe in an angiogram and may explain why the study of [58] showed a higher prevalence of the anomaly compared to the autopsy studies of [12].

The important results to take from these studies are that among the general population, there is a significant percentage which show an incomplete or deficient circle of Willis. The most common variations are hypoplasia of the PCoA's and of the P1 segments of the PCA in the posterior of the CoW, and multiple ACoA's in the anterior. Furthermore, these variations are quite commonly found in combination with one another, but the actual absence of a PCoA is quite uncommon. All such studies will differ slightly in the classification of anatomical variations, based on the population demography examined, the examination method used, and whether or not the individuals were suffering from some type of neurologic dysfunction. In fact, due to the complex demography and the possible variations and combinations which may exist, attempting to classify and obtain statistics on the anatomical variations of the CoW is a difficult task. All of the studies however indicate that the presence of a complete circle of Willis aids in the ability to resist neurologic impairment.

1.5   Blood Vessel Structure

To end this introduction to the cerebral vasculature a brief outline of the structure of blood vessels will be presented, as this information will aid in making certain modelling decisions and also integrate with the outline of the relevant physiology in Chapter 2. As blood is transported from the heart to the tissues the blood vessels undergo repeated branching to progressively smaller sizes and increasing number. Although the blood vessels form a continuum they can be generally classified into *arteries, arterioles, capillaries, venules* and *veins*. For the present study, the focus will be on the delivery of blood *to* the cerebral tissue and not its transport back to the heart via the venules and veins, and as a result only the structure of the first three categories will be outlined. The arteries themselves form a continuum from the largest to the smallest branches, but can also be classified into the larger *elastic* arteries and the smaller *muscular* arteries.

Except for the capillaries and the venules, the blood vessel walls consist of three relatively distinct layers, which are most apparent in the muscular arteries and least apparent in the veins. From the *lumen* to the outer wall of the blood vessels, the layers (or tunics) are the *tunica intima*, the *tunica media* and the *tunica adventitia* (or tunica externa) (Figure 1.14).

The tunica intima consist of a layer of *endothelial cells*, a delicate connective tissue basement membrane, a thin layer of *connective* tissue called the *lamina propria*, and a layer of *fenestrated* (having openings) fibres called the *internal elastic membrane*. The internal elastic membrane separates the tunica intima from the next layer, the tunica media. The tunica media consists of *smooth muscle* cells arranged in a circular fashion around the blood vessel. The tunica media also contains variable amounts of *elastic* and *collagen* fibres, depending on the size of the vessel. An *external elastic* membrane, which separates the tunica media from the tunica adventitia, can be identified at the outer border of the tunica media in some arter-

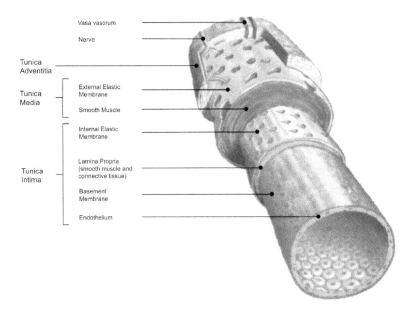

Vasa vasorum

Nerve

Tunica
Adventitia

Tunica
Media

External Elastic
Membrane

Smooth Muscle

Internal Elastic
Membrane

Lamina Propria
(smooth muscle and
connective tissue)

Tunica
Intima

Basement
Membrane

Endothelium

Figure 1.14: The generalized structure of the layers of the blood vessel wall, observable in most vessels except for the capillaries and venules.

ies. A few longitudinally oriented smooth muscle cells occur in some arteries near the tunica intima. The tunica adventitia is composed of connective tissue, which varies from *dense* connective tissue near the tunica media to *loose* connective tissue that merges with the connective tissue surrounding the blood vessels. The relative thickness and composition of each layer varies with the diameter of the blood vessels and its type. The transition from one artery type or from one vein type to another is gradual, as are the structural changes. For arteries and veins greater than approximately 1mm in diameter, nutrients cannot diffuse from the lumen of the vessel to all of the layers of the wall. Nutrients are therefore supplied to their walls by way of small blood vessels called *vasa vasorum*, which penetrate from the exterior of the vessel to form a capillary network in the tunica adventitia and the tunica media. *Sympathetic nerve fibres* innervate the walls of most blood vessels where the nerve terminals project among the smooth muscle cells of the tunica media. Small arteries and arterioles are generally innervated to a greater extent than other blood vessel types.

The afferent and efferent arteries outlined previously, supplying blood to the brain and distributing it to the various cerebral territories, would generally fall under the classification of elastic arteries. These arteries have the largest diameters and the highest blood pressures occurring within them, ranging between *systolic* and *diastolic* values. The tunica intima is relatively thick owing to the large amount of elastic tissue contained within it, although the fibres of the internal and external elastic membranes are typically not recognizable as distinct layers. There is only a small amount of smooth muscle occurring in the tunica intima the tunica adventitia is relatively thin.

The muscular arteries include most of the smaller unnamed arteries and their walls are relatively thick compared to their diameter, mainly because the tunica media contains 25-40 layers of smooth muscle. The tunica intima of these arteries has a well developed internal elastic membrane and the tunica adventitia is composed of a relatively thick layer of collagenous connective tissue that blends with the surrounding connective tissue. The presence of the large amount of smooth muscle allows these vessels to significantly regulate their diameter and thereby the resistance to blood flow and for this reason, the muscular arteries are often termed *resistance* arteries. Muscular arteries can range from 40-300μm where the amount of smooth muscle decreases with the diameter such that those arteries which are in the 40μm diameter range have only approximately three or four layers of smooth muscle in their tunica media.

The arterioles transport blood from the small arteries to the capillaries and are the smallest arteries in which the three tunics can be identified. Arterioles have thick muscular walls and are the in fact primary site of vascular resistance. They range from approximately 40μm to as small as 9μm in diameter. The tunica intima has no observable internal elastic membrane and the tunica media consists of one or two layers of circular smooth muscle cells.

The capillaries are the smallest of a body's blood vessels and most closely interact with the tissue. They consist primarily of a single layer of endothelial cells resting on a *basement membrane.* Outside the basement membrane is a delicate layer of loose connective tissue that merges with the connective tissue surrounding the capillary. Most capillaries range from 7-9μm in diameter and they branch without a change in their diameter. Capillaries are variable in length, but in general they are approximately 1mm in length. Red blood cells flow through most capillaries in a single file and frequently are folded as they pass through the smaller diameter capillaries. Substances cross capillary walls by diffusing through the endothelial cells, through *fenestrae* (small pores in endothelial cells), or between the endothelial cells. Lipid soluble substances, such as oxygen and carbon dioxide and small water soluble molecules readily diffuse through the *plasma membrane*. Larger water soluble substances must pass through the fenestrae or gaps between the endothelial cells. The walls of the capillaries are effective permeability barriers because red blood cells and large water soluble molecules like proteins cannot readily pass through them.

Chapter 2

Physiology

2.1 Introduction

Having outlined the major cerebral arteries and blood vessel structure, the focus of the present chapter is to provide a moderately detailed description of the physiological processes relevant to blood flow in the cerebral vasculature. The chapter will begin by outlining the various biochemical steps in the brain metabolism, continuing with a description of the blood chemistry and its function in meeting the brain's metabolism, the physiology behind the ability of blood vessels to constrict or dilate will be addressed. Finally some clinical mechanics regarding blood flow and pressure will be introduced along with their incorporation into the physiological mechanisms utilized by the body for maintaining homeostasis.

2.2 Brain Metabolism

The brain is one of the most metabolically active organs in the body, utilizing oxygen at a very rapid rate. Since the amount of oxygen stored in the cerebral tissue is small, the brain is absolutely dependent on the continuous replenishment of oxygen by the arterial blood and consciousness will be lost within approximately 10 seconds if blood flow is completely interrupted, leading to irreversible pathological changes within a few minutes [136]. The brain is composed of a variety of tissues and discrete structures that can often function independently or even inversely with respect to one another and as a result, there is a variation in metabolism throughout the brain with typically higher metabolic rates in the grey matter compared to the white matter

for example [136]. This heterogeneity in metabolic rate is further highlighted by the
increase in energy demand in various regions of the brain associated with performing
various tasks such as reading or moving an arm. When considering the brain as a
whole entity however, the overall metabolic rate does not change appreciably when
performing various functions, or even at night during sleep.

The brain does not perform mechanical work like the heart or skeletal muscle,
nor does it perform osmotic work as the kidney does, for example, in concentrating
urine. Instead the major metabolic requirements of the brain are related to its
unceasing electrical activity, requiring energy to actively transport ions across neural
cell membranes in order to sustain and restore the membrane potentials discharged
during the processes of excitation and conduction. The major way in which oxygen
is utilized in the body in order to provide the energy to maintain homeostasis and
sustain life is ultimately through the production of *Adenosine Triphosphate (ATP)*
(often referred to as the 'energy currency' of the body), although some of the oxygen
is also used in a variety of processes relating to the synthesis and metabolism of a
number of neurotransmitters. While there are a variety of metabolic processes in
which ATP can be synthesized from lipids or proteins, in the brain ATP is produced
almost entirely by the oxidation of carbohydrate.

Carbohydrate metabolism in the cerebral tissue essentially involves the combination of glucose and oxygen, producing carbon dioxide, water and ATP. The complete
process involves a level of biochemical complexity that is outside the scope of this
chapter, but the major steps will be outlined as some of the products of those steps
in the reactions can be important when considering the regulation of cerebral blood
flow. The four major steps in order are *glycolysis*, the *citric acid cycle*, the *electron
transport chain* and *ATP phosphorylation* (Figure 2.1) [149]. Glycolysis takes place
in the cytosol of the cells comprising the cerebral tissue and involves a series of reactions (not shown in the figure) by which a molecule of *glucose* is sequentially broken
into two molecules of *pyruvic acid*. During this series of steps a small amount of
ATP is produced from *Adenosine Diphosphate (ADP)* and *Nicotinamide Adenine
Dinucleotide (NAD⁺)* is reduced to *NADH*. NADH is a carrier molecule with two
high energy electrons which may be used at a later stage (the electron transport
chain) to produce ATP, essentially by giving up their energy. The important end
product of glycolysis is pyruvic acid, and depending upon whether oxygen is present
in the cerebral tissue or not, pyruvic acid may follow two different pathways. In the
absence of oxygen *(anaerobic respiration)* NADH is used to convert pyruvic acid to

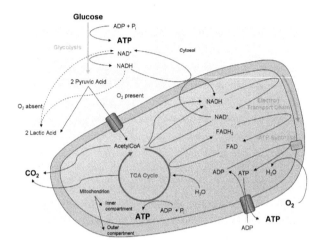

Figure 2.1: The overall processes involved in carbohydrate metabolism as occurs in the cerebral tissue. Aerobic respiration involves the breakdown of glucose into pyruvic acid in the process of glycolysis, pyruvic acid inters the tricarboxylic acid cycle within the mitochondrion of the cell, producing high energy molecules NADH and FADH$_2$ which can be used to synthesize the majority of ATP in the electron transport chain. Anaerobic respiration involves the breakdown of pyruvic acid into lactic acid and a much smaller amount of ATP.

lactic acid (which is transported via the bloodstream to the liver) resulting in an overall production of two molecules of ATP from one molecule of glucose.

In the presence of oxygen *(aerobic respiration)* pyruvic acid enters the inner compartment of the *mitochondrion* (known as the 'power plant' of the cell due to its role in the production of ATP) where it is converted into *Acetyl-CoA* (producing carbon dioxide and reducing NAD$^+$ to NADH) which enters the citric acid cycle, also known as the *TCA* or *Krebs* cycle. The cyclic nature arises because Acetyl-CoA initially reacts with *Oxaloacetic acid* in order to produce *Citric acid,* and after the sequence of reactions in the cycle (not shown in the figure) oxaloacetic acid is one of the end products, so that the sequence may start again. During the cycle NAD$^+$ is reduced to NADH and *Flavin Adenine Dinucleotide (FAD)* is reduced to $FADH_2$ (another high energy electron carrying molecule), both of which enter the electron transport chain. In addition, carbon dioxide and a small amount of ATP are produced. The electron transport chain takes place in the outer compartment of the mitochondrion and involves NADH and $FADH_2$ giving up their electrons to an array of enzymes and electron carriers grouped into several *complexes* in the inner mitochondrial membrane (not shown in the figure). In addition to the NADH formed from the citric acid cycle, NADH in the cytosol (produced during glycolysis) can cause NAD^+ in the inner compartment to be reduced to NADH by passing its electrons through a shuttle molecule in the mitochondrial membrane. As the electron transport process takes place, hydrogen ions are 'pumped' from the inner to the outer compartment, resulting in a higher concentration (lower pH) in the outer compartment. As a result, the hydrogen ions travel down the pH gradient and back into the inner compartment through the enzyme *ATP Synthase* (causing a large amount of ADP to converted into ATP), which combine with electrons from one of the complexes and oxygen (which has diffused from the cytosol into the inner mitochondrial compartment) to form water. The net result from aerobic respiration is that one molecule of glucose can produce thirty six molecules of ATP, most of which are produced during the electron transport chain.

When there is insufficient oxygen for aerobic respiration and the only ATP produced is via glycolysis, the reaction rate speeds up to meet metabolic demand (known as the *Pasteur effect*). However even at its maximum rate the ATP produced from glycolysis is not enough to meet the high energy demands of the brain, further emphasizing the importance of an uninterrupted oxygen supply to the cerebral tissue [136]. Despite the importance of ATP in the brain's metabolism, quantitative measurements of metabolic rate are generally given in terms of oxygen consumption. While these rates differ throughout the brain an average metabolic rate is useful, especially if the cerebral tissue is to be approximated as a homogeneous structure, with a common value being $3.5mL$ of oxygen per $100g$ of brain tissue per minute [136]. Furthermore, under the conditions of aerobic respiration, the stoichiometry of the overall reaction is such that the molar consumption of oxygen and production of carbon dioxide are equal, meaning that the overall cerebral metabolic rate of oxygen consumption may be applied to carbon dioxide production as well.

2.3   Blood

The importance of an uninterrupted oxygen supply to the cerebral tissue has been highlighted in terms of its role in producing ATP and sustaining cerebral function.

It is also important to note that it is not simply the delivery of oxygen which must be maintained for normal metabolism, but also the removal of one of the products of the overall reaction, namely carbon dioxide. Some attention needs to be given as to how these gases are transported to and from the tissues by the blood.

The plasma membranes surrounding cells are permeable to both oxygen and carbon dioxide, which means that they can move freely through the tissues down partial pressure gradients throughout the body (*passive diffusion*), as opposed to requiring a *facilitated diffusion* or *active transport* [149]. To illustrate by example, the partial pressure of oxygen *(oxygen tension)* is higher in the lungs compared to the venous blood, whereas the partial pressure of carbon dioxide *(carbon dioxide tension)* is lower in the lungs. As a result, oxygen taken in by the lungs will diffuse through the alveoli and into the bloodstream, whereas carbon dioxide will diffuse out of the blood to be expelled by the lungs. In the brain the oxygen tension is higher in the blood than in the cerebral tissue, whereas the carbon dioxide tension is higher in the tissue than in the blood. Oxygen therefore moves down a partial pressure gradient into the cerebral tissue while carbon dioxide moves down its partial pressure gradient, into the bloodstream.

The carriage of both of these gases in the bloodstream is a function of the amount of blood flow and also their concentration within the blood. While in water alone the concentration of either gas could simply be related to their partial pressure, the transport molecules and processes that occur within the blood means that oxygen and carbon dioxide transport are part of a complex and intimately linked system. In order to understand the processes, some of the blood chemistry relevant to the carriage of oxygen and carbon dioxide will be covered.

Blood serves a number of functions in the body to help maintain homeostasis, including the transportation of processed and regulatory molecules, maintenance of body temperature, protection against foreign substances, clot formation and the transport of gases, nutrients and waste products. The major component of blood is the *plasma*, which is approximately 91% water, but also contains proteins (such as *albumin, globulins,* and *fibrinogen*), inorganic electrolytes (mainly sodium and chlorine) and serves as a transport medium for nutrients such as glucose and lipids, regulatory substances such as hormones and enzymes, metabolic waste products, as well as gases such as nitrogen, carbon dioxide and a small amount of oxygen.

The other important components of the blood are the *formed elements* which are the red and white blood cells and platelets (*erythrocytes, leukocytes* and *thrombocytes* respectively). Platelets and the white blood cells serve important functions relating to blood clotting and immune responses, whereas the function of the red blood cells is the transport of oxygen, hence making them the most important element for the present discussion. The oxygen carrying capacity of the red blood cells is due to the presence of a protein pigment called *hemoglobin*, which is an assembly of four globular protein subunits, each of which is composed of a protein chain (known as α and β chains) tightly associated with a non-protein *heme* group (Figure 2.2). The importance of hemoglobin lies in the ability of oxygen to bind to the iron atom at the centre of the heterocyclic ring (known as *porphyrin*) of the heme group. Hemoglobin also allows for carbon dioxide to bind to the α amino groups at the ends of the α and β chains, playing a role in the removal of carbon dioxide via the bloodstream.

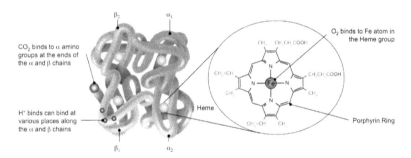

Figure 2.2: The hemoglobin molecule contains four globular protein chains, each tightly coupled with a heme group. The heme group is a porphyrin ring with an iron atom at its centre to which oxygen can bind. In addition other products of metabolism such as carbon dioxide and subsequently hydrogen ions, can bind to the protein chains.

In order to understand the complex coupled system that exists between oxygen and carbon dioxide in the blood it is simplest to first consider how the various forms of each component exist in the blood separately, then combine the overall reactions to illustrate how they interact with one another. Oxygen is transported in the blood by either dissolving into the plasma or combining with hemoglobin. The amount of oxygen dissolved into the plasma is directly proportional to the oxygen tension, which under normal circumstances accounts for only slightly over 1% of the total amount of oxygen transported in the blood (illustrating the importance of hemoglobin). The two α and two β chains of the hemoglobin molecule lie in a 'crumpled ball' (known as the *quaternary structure*) which is of critical importance and governs the reaction with oxygen. The shape is maintained by loose electrostatic bonds between specific amino acids along different chains and also between some amino acids on the same chain. One consequence of these bonds is that the heme groups lie in crevices formed by the electrostatic bonds between the heme groups and *histidine residues*, other than those to which they are attached by normal valency linkages. In un-oxygenated hemoglobin *(deoxyhemoglobin)*, the electrostatic bonds within and between the protein chains are strong, holding the hemoglobin molecule in a tense '*T*' conformation, in which the molecule has a relatively low affinity for oxygen. In oxygenated hemoglobin *(oxyhemoglobin)* however, the electrostatic bonds are weaker and the hemoglobin adopts its relaxed '*R*' state, where the crevices containing the heme groups can open and bind oxygen and the molecule's affinity for oxygen becomes approximately five hundred times greater than in the T state. Binding of oxygen to just one of the four protein chains induces a conformational change in the whole hemoglobin molecule, which increases the affinity of the other protein chains for oxygen. This 'cooperativity' between oxygen binding sites is fundamental to the physiological role of hemoglobin and affects the kinetics of the reaction between hemoglobin and oxygen.

The amount of oxygen transported by hemoglobin depends on the amount of hemoglobin present in the blood and the amount of oxygen actually bound to it (known as *oxyhemoglobin saturation*). Similar to the oxygen dissolved in the plasma, saturation is a function of oxygen tension, but due to the complex kinetics of the chemical reactions between oxygen and hemoglobin, the relationship between oxygen tension and the saturation is nonlinear (Figure 2.3(a)) and its form is of fundamental biological importance [35]. At higher oxygen tensions, as would be the case in the lungs and major arteries, the saturation curve flattens off near one hundred percent. As the blood passes deeper into the tissue, the oxygen tension decreases and so too does the hemoglobin affinity for oxygen, promoting its off loading into the tissue. In addition there are a number of physiological and pathological changes to the blood chemistry which can affect the saturation curve, either promoting the binding of oxygen to hemoglobin or promoting its dissociation.

Some important factors which can effect oxygen saturation are the blood hydrogen ion concentration (pH) and carbon dioxide tension (also affecting blood pH) [55, 95]. The reason for their effect is that hemoglobin can bind protons at various places along the α and β chains, and carbon dioxide at the α amino group, causing

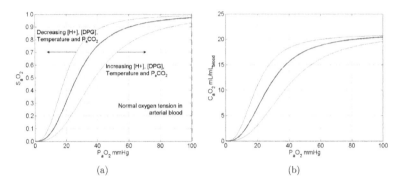

(a)   (b)

Figure 2.3: (a) The oxyhemoglobin saturation curve, illustrating the variation in the
amount of oxygen that binds to hemoglobin as a function of the oxygen tension, and also
the effects of other substances on the position of the curve (b) the actual concentration of
oxygen in the blood stream (including that dissolved in the plasma) as a function of oxygen
tension. Note: the pink dashed line illustrates the normal oxygen tension in arterial blood.

a conformational change in the protein that facilitates the release of oxygen (known
as the *Bohr effect*). Therefore increasing carbon dioxide and hydrogen ion concentration (as would be the case in the cerebral tissue) shifts the saturation curve to
the right, promoting the off loading of oxygen, while decreasing carbon dioxide and
hydrogen ion concentration (as would be the case in the lungs) shifts the saturation
curve to the left, promoting the uptake of oxygen by hemoglobin. Temperature
also affects the saturation curve, but to a much lesser extent than the Bohr effect.
Under hyperthermic or hypothermic conditions the shift in the saturation curve can
become important.

Another important factor affecting the oxygen affinity of hemoglobin is the concentration of the organic phosphate present in high concentrations in the red blood
cells, *2,3 diphosphoglycerate (2,3 DPG)*. 2,3 DPG is an isomer of *1,3 DPG* (which is
one of the intermediaries during glycolysis within the red blood cells) and is synthesized and degraded by *DPG Mutase* and *DPG Phosphatase* respectively, through a
pathway known as the *Rapoport-Luebering shunt*. 2,3 DPG binds to the histidine
and *lysine* residues in the α and β chains of deoxyhemoglobin and acts to stabi-
lize the low oxygen affinity T state, thereby making it harder for oxygen to bind
hemoglobin and more likely to be released to adjacent tissues. As the oxygen tension increases (as would be the case in the lungs) DPG is expelled and the R state

is adopted. 1,3-DPG levels appear to be related largely to cellular ADP and ATP levels [95] and when ATP falls (and ADP rises), a greater proportion of 1,3 DPG is converted through the ATP-producing step and hence greater levels of 2,3 DPG. This mechanism is thus part of a feedback loop to assure a sufficient supply of ATP to meet cellular needs and plays a role in the adaptation to hypoxia.

In contrast to the transport of oxygen in the blood, carbon dioxide can be transported in a number of different forms. It has a much higher solubility in the plasma compared to oxygen, and hence the majority of carbon dioxide is dissolved into the plasma. Only approximately five percent actually remains as dissolved molecular carbon dioxide as it reacts with water to form carbonic acid, with a further dissociation to form bicarbonate ions and hydrogen ions (Figure 2.5) [55, 95]. As a result, approximately ninety percent of the carbon dioxide transported in the bloodstream is in the form of bicarbonate.

Similar to its ability to bind to the α and β chains of the hemoglobin molecule, carbon dioxide can bind to certain amino groups in the plasma proteins, forming carbamic acid. At body pH, the carbamic acid then dissociates almost completely to carbamate and hydrogen ions. In a protein, the amino groups involved in the peptide linkages between amino acid residues cannot combine with carbon dioxide, so the potential for carbamino carriage is therefore restricted to one terminal amino group in each protein chain and to the side chain amino groups that are found in lysine and *arginine*. Since both hydrogen ions and carbon dioxide compete to react with uncharged amino groups, the ability to combine with carbon dioxide is markedly pH dependent. Carbon dioxide also diffuses into the red blood cells where a number of reactions occur, one of which is the formation of carbamino compounds, by the aforementioned binding to hemoglobin. The amount of carbon dioxide transported in the form of carbamino compounds formed with the plasma proteins is usually considered negligible, while the amount formed with hemoglobin is approximately five percent.

Since the reaction of carbon dioxide with water is slow (requiring minutes to achieve equilibrium) the amount transported in the blood in the form of bicarbonate would be much smaller than what is actually observed, given the timescale for gas exchange between the lungs and body tissues. The reaction is catalyzed by an enzyme known as *carbonic anhydrase (CA)*, which is found within the red blood cells and is involved in the transport of carbon dioxide. The accumulation of in-

tracellular bicarbonate and hydrogen ions would rapidly tip the equilibrium of the
reaction against a further dissociation of carbonic acid, but it is circumvented by
two mechanisms. The first is the buffering of hydrogen ions as they bind to the
α and β chains of hemoglobin (the Bohr effect) and the second is the transport of
bicarbonate ions out of the red blood cell, in exchange for chloride ions (to maintain
overall charge neutrality in the cell) through a membrane bound protein known as
Band 3, in what is known as the *chloride shift*. Almost all of the carbon dioxide
transported into the blood stream from the cerebral tissue therefore diffuses into the
red blood cells, accounting for the majority of carbon dioxide being transported as
bicarbonate. Of the total amount of bicarbonate in the blood, approximately thirty
percent remains in the red blood cells, with the remaining seventy percent passing
back into the plasma.

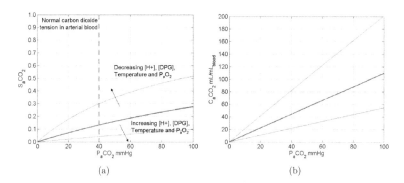

(a)   (b)

Figure 2.4: (a) the carbamino saturation curve illustrating the variation in the amount of
carbon dioxide that binds to hemoglobin as a function of the carbon dioxide tension, and
also the effects of other substances on the position of the curve (b) the actual concentration
of carbon dioxide in the blood stream (including that dissolved in the plasma and that in
the form of bicarbonate) as a function of carbon dioxide tension. Note: the pink dashed
line illustrates the normal carbon dioxide tension in arterial blood.

The amount of carbon dioxide transported in the blood depends upon the carbon
dioxide tension and, similar to oxygen transport, is a non linear function affected by
a number of physiological and pathological changes to the blood chemistry. Similar
to the oxyhemoglobin saturation curve, a *carbamino saturation* curve can be drawn
(Figure 2.4(a)) which is shifted due to changes in the same variables affecting the
oxyhemoglobin saturation curve [35]. Increasing oxygen tension promotes the uptake
of oxygen by hemoglobin and the corresponding dissociation of carbon dioxide (as

mentioned previously when considering oxygen transport) and vice versa, shifting the saturation curve in what is known as the *Haldane effect*. Given that only a small amount of carbon dioxide is actually transported by hemoglobin the most significant changes to carbon dioxide transport are those which affect the levels of bicarbonate, and the major influencing factor is hydrogen ion concentration.

The overall coupled process of oxygen and carbon dioxide transport is illustrated in Figure 2.5, also showing a simplification of the metabolic processes outlined previously. It should be noted that both oxygen and carbon dioxide concentrations vary as a function of the position through the vasculature due to the unloading of oxygen, the uptake of carbon dioxide and their complex interaction with molecules within the red blood cells.

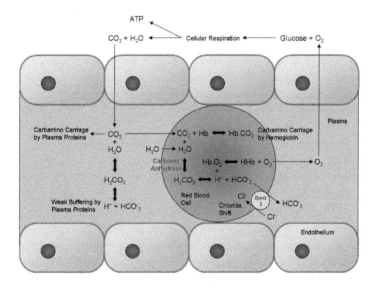

Figure 2.5: The overall processes involved in delivery of oxygen to and the removal of carbon dioxide from the cerebral tissue.

It should also be noted that overall, in the process of transporting the metabolic carbon dioxide out of the tissues and into the lungs, there is only approximately an eight percent increase in the total blood concentration [95]. In addition to this (and despite the earlier argument for uninterrupted blood flow) the amount of oxygen supplied to the cerebral tissue via the arterial blood normally exceeds the metabolic demand such that there is only approximately a thirty percent decrease in total blood oxygen concentration. This proportion, known as the *Oxygen Extraction Fraction* will be shown to be of great importance at a later stage (Chapter 6) in terms of maintaining the oxygen demands of the brain since, although the cerebral tissue cannot survive if blood flow ceases completely, it can tolerate a reduced blood flow by simply utilizing more of the available oxygen.

2.4   Smooth Muscle Contraction

As was illustrated in the previous chapter, blood vessels contain a layer of smooth muscle cells in their tunica media which can either constrict of dilate in order to alter their diameter, thereby changing the resistance to blood flow. While the end goal of this chapter is to provide an outline of the mechanisms which cause this change in arterial diameter and furthermore to link them with the blood chemistry and brain metabolism outlined previously, the present discussion will focus on the physiology of how smooth muscle tone is achieved.

Smooth muscle is one of three types of muscle found in the body, the other two being skeletal and cardiac muscle. Skeletal muscle is a type of *striated* muscle, attached to the skeleton and is used to facilitate movement, by applying force to bones and joints, via contraction. It can generally contract voluntarily, via nerve stimulation, although it can contract involuntarily. Cardiac muscle is also a type of striated muscle, found only in the heart, and causes the contraction used to pump blood throughout the circulatory system. Both skeletal and cardiac muscle can be characterized by the fact that their contractions are rapid and of relatively short duration. Smooth muscle on the other hand, is a type of *non-striated* muscle and can be characterized by the fact that its contractions are much slower but are sustained, *tonic* contractions. Furthermore, smooth muscle contains numerous connections between neighbouring cells known as *gap junctions*, which act as low resistance pathways for the rapid spread of electrical signals throughout the tissue. Smooth muscle is typically classified into *Single* and *Multi-unit* smooth muscle. With single unit smooth muscle the cells are roughly parallel to one another and are gathered together into dense sheets or bands with numerous gap junctions between neighbouring cells. This type of smooth muscle is generally found within the walls

of hollow organs such as the bladder, the uterus, and the gastrointestinal tract, and
is used to move matter within the body via contraction. With multi-unit smooth
muscle the cells are intermingled with connective tissue fibers and have fewer gap
junctions compared to single unit cells. It is the multi-unit smooth muscle which is
typically found in blood vessel walls [149].

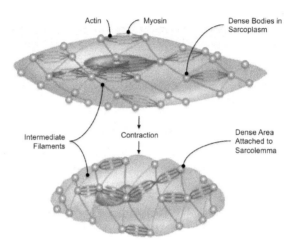

Figure 2.6: A schematic of a single smooth muscle cell, illustrating the important Actin
and Myosin myofilaments and the effect if a contraction on the shape of the cell.

The ability of a smooth muscle cell to contract is due to the presence of *myofil-
ament* proteins, known as *Actin* and *Myosin* (Figure 2.6). Actin myofilaments are
attached to *dense bodies*, which are located within the cell cytoplasm, and to *dense
areas*, which are located within the *sarcolemma* (muscle cell plasma membrane).
Also connected to the dense bodies are the non-contractile *intermediate filament*
proteins, which together with the dense bodies form the intracellular *cytoskeleton*
of the smooth muscle cell, which has a longitudinal or spiral organization and provides the anchoring point such that when the overlap between acting and myosin
increases, the entire muscle cell shortens.

Contraction in smooth muscle can be initiated by mechanical, electrical, or
chemical stimuli. Passive stretching can cause contraction that originates from the
smooth muscle itself and is therefore termed a *myogenic* response. Electrical depolarization of the smooth muscle cell membrane also elicits contraction, most likely

by opening *voltage dependent calcium channels* and causing an increase in the intracellular concentration of calcium. Finally, a number of chemical stimuli such as
Norepinephrine or *Angiotensin II* can cause contraction by binding to specific receptors on the smooth muscle cell plasma membrane (or to receptors on the endothelium
adjacent to the smooth muscle cell), which then leads to contraction.

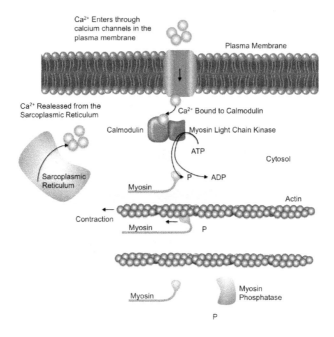

Figure 2.7: The physiological processes involved in a smooth muscle contraction. Note:
the major regulator is intracellular calcium which either enters the cell through channels
in the plasma membrane or is released from internal stores.

The mechanism by which an increase in intracellular calcium stimulates smooth
muscle contraction is illustrated in Figure 2.7. An increase in intracellular calcium
can result from either increased flux of calcium into the cell through *calcium channels*
or by release of calcium from the *sarcoplasmic reticulum (SR)* (an internal calcium
store). The free calcium binds to a special calcium binding protein called *Calmod-
ulin* (the end product denoted *CaM*) which then activates the enzyme *Myosin Light
Chain Kinase (MLCK)*. Myosin light chains are regulatory subunits found on the
myosin which, in the presence of ATP, are phosphorylated by activated MLCK (i.e.

have a phosphate group transferred to them). The myosin light chain phosphorylation leads to *cross-bridge* formation, an attachment between the myosin heads
and the actin filaments involving a conformal change in the shape of the myosin
head, causing it to 'move' along the actin molecule. Relaxation of the muscle occurs
because of the activity of another enzyme called *Myosin Light Chain Phosphatase
(MLCP)*, which removes the phosphate group from the myosin head. If the phosphate group is removed from the myosin head while it is attached to actin then the
cross bridges release very slowly and hence the muscle is able to maintain tension
for a long period of time (known as the *latch state*) [149].

2.5 Clinical Mechanics of Blood Flow

Before starting an outline of the numerous physiological control systems the human
body has for regulating its blood flow, it is important to introduce some clinical
variables pertaining to cerebral blood flow, and how they interact with one another.
While a much more rigorous description of the governing hemodynamic equations
will be given in Chapter 7, these clinical variables can be thought of as simplified lumped parameter relationships, generally considering flowrate and pressure
throughout the brain as a whole.

If the heart was not beating and blood was stagnant within the cardiovascular
system, then blood pressure would still exist because of the tension that the blood
vessel walls exert in order to contain the *blood volume*. The reason that blood flows
throughout the cardiovascular system is that as the heart constricts the volume in
the cardiovascular system is reduced. On its own, this effect would result solely
in the blood pressure increasing, but because of the presence of the heart valves
(allowing for flow in one direction) then the increase in blood pressure will cause
blood to flow through the blood vessels. Blood pressure generated by the beating
of the heart depends in part upon two properties known as *heart rate (HR)* (the
number of contractions per unit time) and the *stroke volume (SV)*(the decrease in
volume of the heart as it constricts and therefore the volume of blood pumped per
beat). Together these two properties form the *cardiac output (CO)*, which is in fact a
flowrate. The other property that affects blood pressure is the *peripheral resistance
(PR)*, which is a complex function of the geometry of the blood vessels (particularly
the diameter) and of the blood itself, affecting how 'difficult' it is for the blood to

flow through the cardiovascular system. The overall relation between these variables
and blood pressure is defined:

$$MABP = HR \cdot SV \cdot PR \qquad (2.1)$$

where $MABP$ is the *mean arterial blood pressure*. Arterial blood pressure *(ABP)*,
varies over the cardiac cycle (as the heart contracts and relaxes) and is in the range
of 120 - 80mmHg between *systole* and *diastole* respectively, for an normal healthy
person. The MABP is hence the time averaged blood pressure over the cardiac
cycle which is approximately 100mmHg for a normal healthy person. As the blood
flows through the cardiovascular system there is a gradual loss in blood pressure
(mainly due to viscous friction) and in addition the pulsatility is dampened out by
the stretch (distensibility) of the blood vessel walls, such that by the time the blood
reaches the venous system the *venous blood pressure (VBP)* is only approximately
5mmHg and non-pulsatile.

The main point of this explanation is that blood pressure in fact is not an
independent input variable into the system, but is the result of the cardiac output
and the peripheral resistance. The reason that this needs to be understood is that
when considering the brain, and applying the form of (2.1), then the relationship
becomes:

$$MABP = CBF \cdot CVR \qquad (2.2)$$

where in this case CBF refers to the *cerebral blood flow* throughout the brain as
a whole and CVR refers to the *cerebrovascular resistance* of the brain to blood
flow. When considering the whole cardiovascular system, any change in peripheral
resistance would cause a change in blood pressure, but when considering solely blood
flow to the brain then changes in CVR do not effect the overall systemic MABP,
but will affect CBF, since these changes either make it 'easier' or 'harder' for blood
to pass through the cerebral vasculature.

Generally speaking, the drive for blood flow through the brain, known as *cerebral perfusion pressure (CPP)* is defined in a similar manner to (2.2) as:

$$CPP = CBF \cdot CVR \qquad (2.3)$$

where in this case CPP is normally defined as the difference between MABP and the *intracranial pressure (ICP)*, although it can also be defined as the difference between MABP and VBP. To expand upon cerebral perfusion pressure it is first necessary to examine the cranial cavity in which the brain is located. Within the $1600mL$ enclosed by the average skull, approximately eighty percent of the volume is occupied by the brain, approximately twelve percent by the blood contained within the cerebral blood vessels and approximately eight percent by the *cerebrospinal fluid (CSF)*. Since the skull is a rigid, liquid filled container, then the increase in the volume of any of the contents must be accompanied by a decrease in the volume of the remaining contents, otherwise a substantial increase in pressure within the cranial cavity will occur.

If the brain enlarges, some blood or CSF must escape to avoid a rise in pressure. If this is not possible there will be a rapid increase in ICP from the normal range of 5-13mmHg to a much higher value. If there is an increase in the volume of either the brain or blood the normal initial response is a reduction in CSF volume within the skull. CSF is forced out into the *spinal sac*, thus ICP is initially maintained. If the pathological process progresses with further increase in volume, venous blood and more CSF is forced out of the skull. Ultimately this process becomes exhausted, when the venous sinuses are flattened and there is little or no CSF remaining in the head. Any further increase in brain volume then causes a rapid increase in ICP. Although CPP is defined in this manner it is important to understand that physically, increases in ICP effectively 'squash' the brain and its blood vessels, increasing the cerebrovascular resistance.

Since the cerebral arteries and arterioles can dilate by relaxing the smooth muscle and thereby decreasing CVR and making it easier for blood to flow through the brain, in doing so they can cause an increase in *cerebral blood volume (CBV)* (the amount of blood within the cranial cavity at any point in time). While it may not be an issue under normal conditions, an increase in ICP (from any one of a number of causes) can mean that the blood vessels may not in fact be able to dilate (or be limited in their capacity to dilate) because of the compression in the brain.

2.6   Mechanisms for the Regulation of Blood Flow

Now that the basic clinical mechanics relating to cerebral blood flow have been outlined, the physiological control systems by which the body regulates cerebral blood flow can be described. As was hinted at previously, there are a number of mechanisms incorporated by the body, which can be broadly divided into mechanisms which regulate blood pressure and those which regulate peripheral resistance. Furthermore, these mechanisms may be categorized in terms of their timescale into those which act over a matter of seconds or those which act over days to weeks, as well as whether or not they act globally on the whole cardiovascular system or are specific to a single organ. Not all of these mechanisms are completely relevant for the present study in that they will not constitute any part of the mathematical model applied to the circle of Willis geometries, but at least a brief description will be given, firstly for the purposes of understanding the complete picture and secondly as their description can aid in the justification of their omission from the model.

The regulation of blood pressure can be categorized into both short term and long term mechanisms. Mechanisms acting in the short term tend mainly to involve the autonomic nervous system and utilize feedback loops between the brain, heart and blood vessels. Through these feedback loops, the heart rate, stroke volume and peripheral resistance can all be altered, so that the blood pressure will change in accordance. The major mechanisms are arterial *baroreceptors* and *chemoreceptors*, the *adrenal medullary mechanism* and the *central nervous system (CNS) ischaemic response*.

The arterial baroreceptor reflex (Figure 2.6) is initiated by nerve endings located in the wall of each internal carotid artery at the carotid sinus and in the wall of the aortic arch. Changes in blood pressure affect the stretch on the receptors and subsequently the action potential firing rate through the nervous system. The signals enter the medulla oblongata causing a feedback to the heart altering HR and SV as well as to the peripheral circulatory system to alter PR. An increase in ABP causes an increased baroreceptor firing rate, resulting in a reduced HR, and SV as well as a reduced PR to restore blood pressure, and vice versa. The aortic baroreceptors operate, in general, at pressure levels about 30 mmHg and above and are most sensitive around their normal operating blood pressure. They respond extremely rapidly to changes in pressure (even over the cardiac cycle) but are thought to be

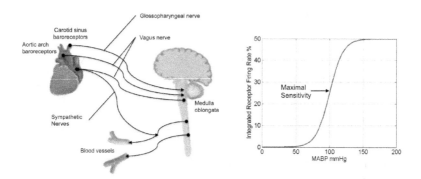

Figure 2.8: The baroreceptor reflex involves a feedback from the receptors at the aortic arch and carotid sinus, to the medulla oblongata, which in turn alters heart rate and stroke volume as well as blood vessel tone. Note: the baroreceptor reflex has maximal sensitivity about small deviations from the normal blood pressure, but adapts this normal level over a period of days.

of little long term importance as they reset within days to whatever pressure they are exposed to.

Closely associated with the baroreceptor system is a chemoreceptor reflex that operates in much the same way as the baroreceptor reflex except that chemoreceptors are sensitive to oxygen, carbon dioxide, or hydrogen ions. Whenever the arterial pressure falls below a critical level, the chemoreceptors become stimulated because of diminished blood flow and therefore diminished availability of oxygen as well as excess buildup of carbon dioxide and hydrogen ions that are not removed by the slow flow of blood. Signals transmitted from the chemoreceptors to the medulla oblongata result in a feedback to the heart to increase HR and SV as well as to increase PR. The chemoreceptors are not stimulated strongly by pressure changes until the arterial pressure falls below 80 mmHg. It is hence at the lower pressures that this reflex becomes especially important to help prevent further pressure drops.

The adrenal medullary mechanism is activated when stimuli result in a substantial increase in sympathetic stimulation of the heart and blood vessels. Cells in the adrenal medulla synthesize and secrete the *catecholamines* known as *epinephrine (adrenaline)* and *norepinephrine (noradrenaline)*. Following their release into blood, these hormones bind adrenergic receptors on target cells, where they induce essentially the same effects as direct sympathetic nervous stimulation; increased heart rate, stroke volume and constriction of the blood vessels.

The central nervous system ischemic response is initiated when blood flow to the vasomotor centre in the lower brain stem becomes decreased enough to cause the neurons in the medulla oblongata to respond directly to the ischemia and become strongly excited. It is thought that the buildup of carbon dioxide causes the response (although lactic acid or other acidic substances may be important too) and when this occurs, the systemic arterial pressure often rises to a level as high as the heart can possibly pump. The CNS ischemic response is one of the most powerful of the sympathetic vasoconstrictor system. However, it does not become significant until the arterial pressure falls far below normal, down to 60 mmHg and below, reaching its greatest degree of stimulation at a pressure of 15 to 20 mmHg. Therefore, it operates principally as an emergency arterial pressure control system that acts rapidly and powerfully to prevent further decrease in arterial pressure whenever blood flow to the brain decreases dangerously close to a lethal level.

Mechanisms for blood pressure regulation in the long term tend to occur mainly through hormonal mechanisms, as opposed to mainly nervous mechanisms in the short term. These hormones affect the functioning of the kidneys, altering the total blood volume and hence the blood pressure. The important mechanisms which act through the kidneys are the *Renin-Angiotensin-Aldosterone* mechanism and the *Vasopressin* mechanism, but there are in addition, two other important mechanisms which do not utilize the kidneys; the *Fluid-Shift* mechanism and the *Stress-Relaxation* response.

The renin-angiotensin-aldosterone system plays an important role in the regulation of blood pressure via blood volume. The most important site for renin release is the kidney, where sympathetic stimulation, renal artery hypotension, and decreased sodium delivery stimulate the release of the enzyme *Renin* by the kidney, which converts *angiotensinogen* into *angiotensin I*. Angiotensin I is then converted to *angiotensin II (Ang II)* by *angiotensin-converting enzyme (ACE)*, which is found mainly in lung capillaries. Ang II circulates in the blood causing vasoconstriction and hence increases PR, but also acts on the adrenal cortex causing the release of *aldosterone*, which in a circular fashion, acts on the kidneys to increase water reabsorption and hence blood volume.

The Vasopressin (or *Antidiuretic Hormone, ADH*) mechanism works together

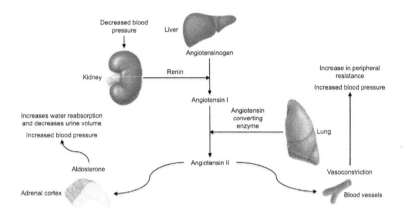

Figure 2.9: The Renin-Angiotensin-Aldosterone mechanism regulates blood pressure by the alteration of blood volume and level of blood vessel constriction. The mechanism acts in the long term via the release of hormones from the kidney and liver.

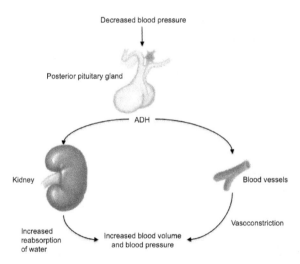

Figure 2.10: The Antidiuretic Hormone mechanism regulates blood pressure via the alteration of blood volume and the level of blood vessel constriction. The mechanism acts in the long term via the release of hormones from the pituitary gland in the brain.

with the renin-angiotensin-aldosterone system in response to changes in blood pressure. The stimulus is via the firing rate of the baroreceptors in response to the changes in blood pressure, which initiate the release of ADH from the *posterior pituitary gland*. The primary function of ADH in the body is to regulate extracellular fluid volume by affecting renal handling of water. ADH acts on renal collecting ducts to increase water permeability, which leads to decreased urine formation. This increases blood volume, cardiac output and arterial pressure. ADH also regulates blood pressure by causing increased vasoconstriction.

The fluid shift mechanism begins to act within a few minutes but requires hours to achieve its full functional capacity. It plays a very important role when dehydration develops over several hours, or when a large volume of saline is administered over several hours. The fluid shift mechanism occurs in response to small changes in pressures across capillary walls. As blood pressure increases, some movement of fluid into the interstitial space helps prevent the continuing development of high blood pressure. As blood pressure falls, interstitial fluid moves into capillaries, which counteracts a further decline in blood pressure. The fluid shift mechanism is a powerful method through which blood pressure is maintained because the interstitial volume acts as a reservoir in equilibrium with the large volume of intercellular fluid.

The stress-relaxation response is characteristic of smooth muscle cells. When blood volume declines, blood pressure also decreases, causing a reduction in the force applied to smooth muscle cells in blood vessel walls. As a result, during the next few minutes to an hour, the smooth muscle cells contract, reducing the volume of the blood vessels and thus resisting a further decline in blood pressure. Conversely, when blood volume increases rapidly, such as during a transfusion, blood pressure increases and smooth muscle cells of the blood vessel walls relax, resulting in a more gradual increase in blood pressure. The stress relaxation mechanism is most effective when changes in blood pressure occur over a period of many minutes.

With the mechanisms outlined thus far, peripheral resistance has been altered (in combination with heart rate and stroke volume) in order to bring about a change in blood pressure. This is the case because these mechanisms tend to act globally throughout the entire cardiovascular system (e.g. increasing heart rate and stroke volume affects blood pressure everywhere in the body). When considering an individual organ such as the brain however, changes in its resistance will not affect the overall systemic blood pressure significantly, but will affect the amount of blood flow through that organ. This 'local regulation' of blood flow is known as *autoregulation* and is defined as the intrinsic ability of an organ to maintain a relatively constant blood flow despite changes in perfusion pressure, within a certain physiological blood pressure range (Figure 2.11). Autoregulation has a very strong influence in certain organs such as the heart and the kidney and of course the brain, where it is known as the *cerebral autoregulation mechanism*.

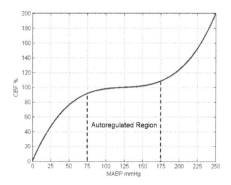

Figure 2.11: A standard cerebral autoregulation curve, illustrating the relatively constant level of blood flow over a range of cerebral perfusion pressures. The autoregulated region involves the vascular smooth muscle altering its tension and hence the caliber of the blood vessels, thereby altering the overall cerebrovascular resistance to blood flow. Outside of the autoregulated region (i.e. at very high or low blood pressures) where the smooth muscle is maximally constricted or dilated respectively, the blood flow is then a function of the blood pressure.

There are a number of mechanisms and chemical pathways encompassed by cerebral autoregulation to bring about the overall autoregulation curve illustrated in Figure 2.11 and to date, the complete pathways and their interactions have not yet been fully elucidated. Even so it is common to broadly classify autoregulation into *metabolic, myogenic, neurogenic* and *shear stress* components. Due to the complexity only a brief introduction will be given to the various components, outlining the more commonly proposed mechanisms. It is important to note that despite the complexity and variety of the chemical pathways involved, essentially all of them result in either alterations in the concentration of intracellular calcium in the vascular smooth muscle (VSM) cell, or alterations in the sensitivity of MLCK to calcium.

Myogenic autoregulation is generally defined as the ability of the smooth muscle cells to respond directly to changes in blood pressure. The force from the blood pressure that is applied to the endothelium is transmitted through the cytoskeleton and extracellular matrix to proteins in the plasma membrane known as *Integrins*, creating a tensile force on the smooth muscle cell. Changes in this force cause rapid architectural distension which then brings about a series of chemical events at the sarcolemma, leading to the alterations in the muscle tone. An increase in blood pressure causes an increase in the tension applied to the VSM and results in constriction so that there is an overall reduction in the diameter of the blood vessel, thereby reducing the applied tension. Conversely a reduction in blood pressure leads to a reduced tension applied to the VSM, a relaxation of cell and an increase in arterial diameter, thereby stabilizing the applied tension.

One of the proposed transduction mechanisms in response to an increase in pressure is the opening of *non-selective stretch activated cationic channels (SAC)* [39, 59, 134], which causes influx of cations such as sodium, potassium and calcium (Figure 2.6). This influx leads to *membrane depolarization* (the membrane potential brought about by the imbalance of cations between the extracellular space and cytosol is reduced as cations flow in), which induces the opening of *voltage gated calcium channels* (Ca_{VG}^{2+}) and an influx of calcium, leading to a rise in intracellular calcium levels and hence muscle tone. The increase of intracellular calcium from this pathway also involves hydrolization of the phospholipid known as *Phosphatidylinositol bisphosphate (PIP$_2$)* by the enzyme *Phospholipase C (PLC)* into two important *second messengers* known as *Inositol Triphosphate (IP$_3$)* and *Diacylglycerol (DAG)*. It is known that IP_3 diffuses into the cytosol and stimulates the release of calcium from the SR and may therefore play a part in the overall rise in intracellular calcium levels. DAG on the other hand remains associated with the sarcolemma, but given a cytosolic increase in calcium can activate the enzyme known as Protein Kinase C (PKC) which can directly phosphorylate the myosin light chains (similar to MLCK) promoting muscle contraction. Another possible pathway involved in the myogenic response is the release of the *omega-6 fatty acid* known as *Arachidonic Acid (AA)* via the membrane protein known as *Phospholipase A2 (PLA$_2$)* in response to distension of the plasma membrane . It is thought that AA and its metabolites in the *Cytochrome P450 pathway*, such as *20-Hydroxyeicosatrienoic Acid (20-HETE)* and *Epoxyeicosatrienoic Acid (EET)* can inhibit calcium sensitive potassium channels (K_{Ca}^+) which would maintain depolarization of the VSM cell and thereby potentiate myogenic constriction [39, 47, 134]. Furthermore it is also thought [39] that AA can inhibit MLCP, thereby enhancing calcium sensitivity. One of the final possible pathways to be mentioned here is the binding of the *small GTPase* protein *RhoA* to the membrane protein *caveolin-1 (CAV-1)* in response to distension of the plasma membrane. This binding causes the activation of the actin binding enzyme *Rho Kinase (RhoK)* which can inhibit MLCP and thereby increase calcium sensitivity [59, 60, 167]. In addition RhoK activation has a key role in cytoskeletal rearrangement via the formation of Actin intermediate filaments, as a longer term response to changes in blood pressure (not shown in the figure). Despite its quick response to variations in blood pressure, the myogenic mechanism is also generally thought to play the major role in the development of a basal level of blood vessel tone, which vasodilators can reduce. Furthermore, there is evidence to illustrate the importance of *Gap Junctions* (connections between cells which allow for certain molecules to pass through them) in the myogenic response [87], where the depolarization or secondary messenger effects brought about by changes in blood pressure can be propagated along the tunica media of blood vessels.

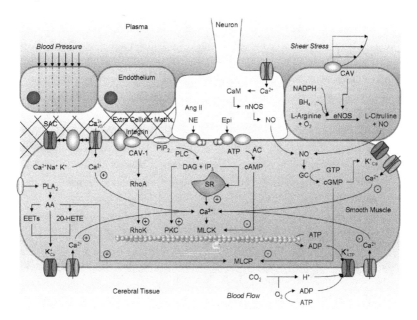

Figure 2.12: The more commonly proposed mechanisms involved in the various components of the cerebral autoregulation mechanism. Note: the ⊕ symbol indicates a pathway which promotes an increase in muscle tone and the ⊖ symbol indicates a pathway which promotes a decrease in muscle tone.

Nitric Oxide (NO) is now accepted as playing an extremely important part in the regulation of blood flow in the cerebral vasculature [108] and one pathway by which it is released is via the shear stress component of cerebral autoregulation. The effect of shear stress on cerebral autoregulation is a more recent discovery compared to the myogenic or metabolic components and occurs through the effect of friction from the blood against the endothelium as it flows through the blood vessels. The mechanism by which this effect is thought to occur begins with the mechanical force from the shear stress activating (through a series of pathways not completely elucidated, but involving the caveolae) the enzyme *Endothelial Nitric Oxide Synthase (eNOS)*. In the presence of other *cofactors* (substances necessary to the function of an enzyme) *tetrahydrobiopterin (BH_4)* and *Nicotinamide Adenine Dinucleotide Phosphate (NADPH)*, eNOS causes conversion of the amino acid *L-Arginine* and oxygen to *L-Citrulline* and NO [38]. NO can diffuse readily throughout tissues and moves into the cytosol of the smooth muscle cell, where it activates the enzyme *Guanylate Cyclase (GC)*, causing the conversion of *Guanosine Triphosphate (GTP)* to *Cyclic Guanosine Monophosphate (cGMP)*. It is thought that cGMP can affect smooth muscle tone in two ways; the first via the interaction with calcium specific potassium channels (K_{Ca}^+), leading to a *membrane hyperpolarization* (the membrane potential brought about by the imbalance of cations between the extracellular space and cytosol is increased) and a closing of Ca_{VG}^{2+} channels, leading to a reduction in intracellular calcium and hence reduced muscle tone [14]. The second way in which cGMP is thought to lead to dilation is through its stimulation of MLCP, breaking cross-bridges and thereby reducing tone [22]. It is starting to be proposed that the myogenic and shear stress NO components of autoregulation are in constant opposition [47], where NO induced vasodilation attenuates the myogenic induced constriction.

Despite the fact that cerebral blood vessels are strongly innervated [148], the role of the nervous system in cerebral autoregulation is not well defined. There are a number of pathways leading to changes in VSM tone however, two of which involve the *catecholamine neurotransmitters Norepinephrine (NE)* and *Epinephrine (Epi)*. NE released from the *presynaptic terminals* of neurons can act via *a1-adrenoceptors* to activate PLC causing the hydrolization of PIP_2 into IP_3 and DAG as outlined previously within the myogenic component of autoregulation. The same pathway is then used to stimulate the release of calcium from the SR and phosphorylate the myosin light chains through PKC promoting muscle contraction (note that Ang II acting as part of the Renin-Angiotensin-Aldosterone mechanism for the regulation of blood pressure utilizes the same pathway). Epi on the other hand acts through

β_2-adrenoreceptors to stimulate the enzyme *Adenylyl Cyclase (AC)* to convert ATP to *Cyclic Adenosine Monophosphate (cAMP)*. cAMP is an inhibitor of MLCK and therefore this pathway promotes a reduction in MLCK phosphorylation and hence reduces VSM tone. It is important to note that these pathways can also be activated by NE and Epi in the blood stream as part of other mechanisms such as the adrenal medullary mechanism. Another important neurotransmitter of more recent discovery is NO [148]. The mechanism by which NO is formed in the presynaptic terminal is essentially the same as that for the endothelium [44] except that rather than being released from the caveolae, calcium enters the terminal through calcium channels as a result of an *action potential* (note: both NE and Epi are also released by via action potentials), and the enzyme used in its conversion is *Neuronal Nitric Oxide Synthase (nNOS)* as opposed to eNOS. Once produced, NO diffuses into the VSM where it activates GC and follows the same pathway as described previously for endothelial NO, promoting a reduction in intracellular calcium levels and stimulating MLCP, leading to reduced VSM tone. The link between the action potentials stimulating these neurogenic pathways and parameters relevant to cerebral autoregulation such as blood flow and pressure is currently not well defined. It is likely however that the innervation of the blood vessels and these pathways may not play such an important role in terms of the highly local regulation of CVR, but may function as part of other mechanisms such as the baroreceptor and chemoreceptor reflexes for example.

It has been well accepted that alterations in the chemicals involved in metabolism *(metabolites)* cause vasodilation and hence alterations in CVR, where oxygen, carbon dioxide, hydrogen ions, and adenosine have been identified for their importance. The accumulation of these metabolites is related to the delivery of oxygen to, and the removal of carbon dioxide from, the tissue which is a function of the cerebral blood flow. Carbon dioxide is constantly being produced as part of the carbohydrate metabolism occurring in the cerebral tissue and its removal is dependent on the amount of blood flow. Given a reduction in blood flow, then the concentration of carbon dioxide will start to build up, although it is not molecular CO_2 itself which is considered important in terms of vasodilation, but rather the fact that it dissociates into carbonic acid, causing an increase in hydrogen ion concentration in the tissue (a drop in pH) [86, 147]. There have been numerous studies illustrating that the drop in extracellular pH (i.e. external to the VSM cell) [7] affects ATP specific potassium channels (K_{ATP}^+) on the plasma membrane causing a hyperpolarization of the cell and hence a reduced calcium influx through Ca_{VG}^{2+} channels [64, 65, 72, 123]. This reduced calcium influx leads to decreases in intracellular calcium levels and hence a reduction in muscle tone. Metabolism also requires the constant delivery of oxygen to synthesize ATP and when its delivery is impeded there is an increase in the ratio of ADP to ATP as it cannot be synthesized rapidly enough. In this case there is evidence to show that the opening of K_{ATP}^{+} channels is reduced, also leading to a hyperpolarization of the cell and a reduced calcium influx, hence vasodilation [20]. Furthermore it is possible that given the condition where there is not enough oxygen present for aerobic respiration to occur, the increase in lactic acid as a byproduct of anaerobic respiration may also lead to a drop in pH, leading to a hyperpolarization via an interaction with K_{ATP}^{+} channels.

One emerging idea is that the metabolic component of autoregulation is most effective in the smallest resistance arteries and does not in fact play a major role in the overall alteration of CVR [47]. Instead it is hypothesized that the dilation induced in these smaller arteries and arterioles from the buildup of metabolites increases the shear stress further upstream in the larger arteries, initiating a NO dependent vasodilation, thereby altering CVR to a much more significant degree (termed *conducted dilation*). This hypothesis is enhanced by a number of studies which have shown that inhibiting the formation of NO tends to attenuate the vasodilator response to changes in carbon dioxide and hydrogen ions, but does not completely abolish it [72, 123]. Furthermore it is hypothesized that the myogenic response is also much stronger in the larger arteries and in addition to its role of maintaining a basal tone (in constant competition with the NO dependent vasodilation) can play an important role in protecting the downstream capillaries by its constriction in response to high blood pressure.

It is clear that there is great complexity in the chemical pathways involved in cerebral autoregulation and their interaction. It should be noted that this discussion has only attempted to outline some of the more commonly proposed mechanisms and combine them in order provide a basic outline of the overall picture. It is likely that in years to come, vasoconstrictor and vasodilator pathways will be expanded upon and their interactions more clearly defined, new pathways may be found, and perhaps commonly existing pathways will be either completely dismissed, or found to be of less importance than was originally thought. In terms of mathematical modelling, the heterogeneity between different levels of the vasculature and the as yet incomplete pathways mean that attempting to model this level of complexity would be an insurmountable task, and will not be attempted for the present study.

But this discussion should keep the assumptions made with any model in perspective and provide a good starting point for future improvements.

Chapter 3

Stroke

3.1 Introduction

The previous two chapters have focussed on the anatomy and physiology relevant to cerebral hemodynamics, and how to a certain extent, through the anastomotic connections of the circle of Willis and the cerebral autoregulation mechanism, both attributes can maintain oxygen delivery to the cerebral tissue to meet its metabolic demands. With this information outlined, it is time to introduce the neurologic dysfunction known as stroke, which is the third leading cause of death in many developed countries, following cardiovascular disease and all forms of cancer, and is the largest cause of long term adult disability [5]. A stroke, or 'brain attack', is caused by an acute interruption in the blood supply to the brain, and while there are a number of pathological causes for this interruption, two major classifications for stroke type can be identified. The first category of stroke occurs when a blood vessel in the brain breaks, causing blood to leak into the brain tissue or cerebrospinal fluid surrounding the brain, and is known as a *hemorrhagic* stroke. The second type occurs due to a blockage in a blood vessel perfusing a region of the brain, such that the blood supply is interrupted, and is known as an *ischaemic stroke.*

When either of these two types of stroke occur the brain tissue begins to die, leaving a region of *infarct* (dead) tissue where irreversible brain damage has occurred. With ischemic stroke however, a region known as the *ischemic penumbra* can also be identified where the damage is not necessarily permanent and the functioning of this region may be restored if pharmacological intervention can restore oxygen supply in a timely manner. When brain cells die during a stroke, the abilities controlled by that area of the brain such as speech, movement, and memory are lost.

How a stroke patient is affected depends on where the stroke occurs in the brain and the extent to which the brain is damaged. For example, an individual suffering a small stroke may experience only minor problems such as weakness of an arm or leg. People who have larger strokes may be paralyzed on one side or lose their ability to speak, or the stroke may be fatal.

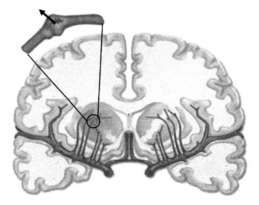

Figure 3.1: Hemorrhagic stroke occurs when blood from a burst artery is forced into either the brain tissue, or into the narrow space between the brain surface and the layer of tissue that covers it.

3.2   Hemorrhagic Stroke

Hemorrhagic stroke (Figure 3.1) is caused by a hemorrhage (a breakage) of a blood vessel in the brain and accounts for approximately 12% of all strokes. Hemorrhages can be caused by a number of disorders which affect the blood vessels, including long standing hypertension and cerebral aneurysms. An aneurysm is a weak or thin spot on a blood vessel wall, and the force of the blood pressure causes it to 'bulge' out from the blood vessel. The tendency to form aneurysms can be inherited, and generally they develop over a number of years, not usually causing detectable problems until they break. There are two types of hemorrhagic stroke, classified as *intracerebral hemorrhage* and a *subarachnoid hemorrhage*, based on where the breakage occurs. Intracerebral hemorrhage occurs when a blood vessel within the brain ruptures, causing bleeding into the brain tissue itself. Intracerebral hemorrhage usually occurs in selected parts of the brain, including the basal ganglia, cerebellum, brainstem, or cortex. Subarachnoid hemorrhage occurs when a blood vessel bursts in a large artery on or near the thin, delicate membrane surrounding the brain. Blood spills into the area around the brain, causing the brain to be surrounded by blood contaminated fluid. In either case the sudden increase in pressure within the brain can cause damage to the brain cells surrounding the blood. If the amount of blood increases rapidly, the sudden buildup in pressure can lead to unconsciousness or death.

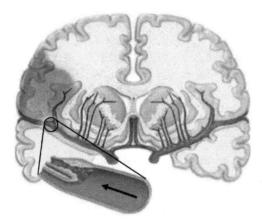

Figure 3.2: Ischaemic stroke occurs when an artery in the brain becomes blocked and the supply of blood to a region of brain tissue is cut off.

3.3 Ischemic Stroke

Ischemic stroke (Figure 3.2) is by far the most common form of stroke, accounting for approximately 88% of all strokes, where the blockage to an artery can occur via thrombosis or embolus. A thrombus is a built up blood clot which usually forms around atherosclerotic plaques and gradually narrows the lumen of the artery, impeding blood flow to distal tissue. Since blockage of the artery is gradual, onset of symptomatic thrombotic stroke is slower. A thrombus itself can lead to an embolic stroke if the thrombus breaks off, whereby it is termed an embolus. An embolus is most frequently a blood clot, although it can also be a fragment of atherosclerotic plaque detached from a blood vessel wall, or a number of other substances (e.g. fat globules). The key feature of the embolus is that it travels in the bloodstream until it becomes wedged in an artery, cutting off its blood supply. Because an embolus arises from elsewhere, local therapy only solves the problem temporarily, thus the source of the embolus must be identified. The onset of the embolic blockage is sudden, resulting in symptoms that are usually maximal at start. Also, symptoms may be transient as the embolus lyses and moves to a different location or dissipates altogether.

The classification of ischemic stroke subtypes is complex and has received considerable study over the years, but it is generally accepted that the major source of ischemic stroke is from embolism, and hence the distinction between thrombotic and embolic stroke somewhat out dated. One of the current classification systems is the *Trial of Org 10172 in Acute Stroke Treatment* (or *TOAST*) classification system which bases the subtypes on their etiology (i.e. their cause) [8]. The five categories postulated in the TOAST system are large artery atherosclerosis, cardioembolism, small artery occlusion (lacune), stroke of other determined etiology, and stroke of undetermined etiology. Large artery atherosclerosis involves the buildup of atherosclerotic plaques in the arterial wall, causing a narrowing of the vessel diameter (stenosis) and if the arteries become too narrow, blood cells may collect and form blood clots. These blood clots can block the artery where they are formed (thrombosis), or can dislodge and become trapped in arteries deeper in the brain (embolism). This category includes atherosclerosis in the larger afferent arteries (where stenosis or occlusion at the carotid bifurcation is common) as well as the major cerebral arteries and their cortical branches. Cardioembolism involves arterial occlusions which are presumed to arise due to an embolus originating from the heart. These emboli can occur as a result of atrial fibrillation (an irregular heart rhythm) where the erratic motion of the atria leads to blood stagnation and potentially clot formation, from a heart attack, or abnormalities of the heart valves, as well as other causes. The emboli travel into the brain where they become lodged in an artery. Small artery occlusion takes place in the smaller deep penetrating arteries, primarily the lenticulostriate branches of the MCA, but also the branches arising from the ACA, AChA, PCA and BA. The term lacune refers to the small deep infarct regions within the cerebral tissue, generally with a maximum diameter of 1.5cm. The occlusion may be a result of atherosclerotic plaque buildup in these arteries or from embolism. Occlusion is most commonly caused by a process of concentric thickening of the penetrating blood vessel wall primarily due to the effect of longstanding high blood pressure (known as *liphyalinosis*), but can also be due to *atheromatous plaque* from the parent vessel occluding the ostium (origin) of the penetrating blood vessel, and in rarer occasions may even be due to small emboli.

A significant variation exists in the reported survival rates for patients with stroke and its several subtypes. The two major factors potentially contributing to this variability include methodological differences between studies and a decline in fatality rates over time, partly due to improved supportive care. The study of Matsumoto et al [98] involved the individuals among the general population of Rochester, Minnesota who had suffered a stroke between 1955 and 1969. The results showed that short term (thirty day) survival rates are substantially higher for patients who have sustained an ischemic rather than a hemorrhagic stroke. In the study of Kelly-hayes et al [78], a total of 213 patients with strokes occurring between 1971 and 1981 were evaluated. Patients in whom stroke was suspected hospitalized and seen by a neurologist. Almost three quarters of the patients (72%) survived more than thirty days. Acute survival was dependent on stroke type and was negatively influenced by severity of neurologic impairment and age. Only about one third of hemorrhagic stroke patients survived the acute phase, but more than three quarters of those with ischemic stroke (cerebral thrombosis or cerebral embolism) were alive at the twenty day mark. Despite the increased mortality from hemorrhagic stroke, it is much less common, therefore the focus of the present study is geared toward modelling cerebral perfusion as a clinical tool for the prediction and prevention of ischaemic stroke. In terms of the TOAST classification system the mortality is higher among patients with large artery atherosclerosis than among patients with lacunes [130]. Recurrent strokes are more likely among patients with cardioembolic stroke than among patients with stroke of other causes [84]. The one-month mortality after cardioembolic stroke is also higher than that with strokes of other etiologies [21].

Relating the effects of ischaemic stroke back to the various cerebral arteries and the regions of the brain which they supply can give an indication of the neurological damage which is likely to result from a reduction in blood supply. Ischemic strokes in the anterior circulation (including both the ACA and MCA) are caused most commonly by occlusion of one of the major *intracranial* arteries or of the small single *perforator* arteries. The most prevalent causes of arterial occlusion involving the major cerebral arteries are emboli, predominantly from large vessel atherosclerosis or cardioembolism.

The ACA supplies blood to the frontal lobe (including Broca's area and the precentral gyrus) and the postcentral gyrus of the parietal lobe, the anterior of the corpus callosum, and the anterior diencephalon. Occlusion of the ACA is uncommon, occurring in only 2% of cases [2], however atherosclerotic deposits can occur in the proximal segment of the ACA leading to major motor and somatosensory deficits. These may include paraplegia affecting the lower extremities, incontinence, personality change, and contralateral grasp reflex. Also, if the left hemisphere is involved then language deficits are almost invariably found as Broca's area will be damaged. The AChA supplies the lateral thalamus and posterior region of the internal capsule. Occlusion of the AChA occurs in fewer than 1% of anterior circulation strokes. Ischemia in the OPhA resulting from ICA occlusion is often transient in nature, with transient monocular blindness occurring in approximately 25% of patients. Central retinal artery ischemia however is relatively uncommon, presumably because of the efficient collateral supply.

Occlusion of the MCA or its branches is the most common type of anterior circulation infarct, accounting for approximately 90% of infarcts and two thirds of all first strokes [2]. Of MCA territory infarcts, 33% involve the deep MCA territory, 10% involve both superficial and deep MCA territories, and over 50% involve the superficial MCA territory. The MCA supplies a large region of the brain including the internal capsule, caudate nucleus, and globus pallidus, as well as most of the temporal lobe (including Wernicke's area), anterolateral frontal lobe, parietal lobe (including the postcentral gyrus) and the insula. Occlusion in the proximal M1 segment will also lead to motor and somatosensory deficits and may cause contralateral hemiplegia affecting the face, arm and leg, homonymous hemianopia (a defect in the visual fields), and aphasia (an impairment in understanding and/or formulating complex, meaningful elements of language). An occlusion of the distal M3 branches that supply the deep white and grey matter (usually embolic in nature) can produce a lacunar type of stroke accompanied by specific neurologic deficits. These occlusions account for as many as 20% of ischemic strokes.

Occlusion of the vertebrobasilar system accounts for approximately 20% of ischaemic stroke [2] and the mortality of patients with basilar artery occlusion is high (75 - 80%). Most survivors of BA occlusion have severe persisting disability. The most common vascular condition affecting the vertebrobasilar system is atherosclerosis, in which plaques cause narrowing and occlusion of the large vessels. Occlusions of the small vessels lead to lacunes, which may appear as single or multiple regions scattered widely throughout the subcortex and brainstem. There is high mortality associated with vertebrobasilar occlusion because it perfuses a number of areas vital for survival, such as the medulla oblongata, pons, and midbrain, as well as the cerebellum, thalamus, and occipital cortex. Patients with small infarctions usually

have a benign prognosis with reasonable functional recovery. Infarctions in the vertebrobasilar system have some characteristic clinical features that distinguish them
from those in the hemispheres including ataxia (loss of the ability to coordinate
muscular movement), dysphagia (difficulty in swallowing or inability to swallow),
vertigo, nausea, vomiting and visual field deficits. In contrast to infarction in the
cerebral hemispheres however, aphasia and cognitive impairments are absent.

In general, patients with PCA distribution strokes exhibit less overall chronic
disability than those with anterior cerebral, middle cerebral, or basilar artery infarctions and occlusion of the PCA only occurs in approximately 5 - 10% of ischaemic
strokes [2]. The PCA supplies the occipital lobe (including the primary visual cortex,
so its occlusion may also lead to visual field losses. Other clinical symptoms include
colour blindness, failure to see movements, verbal dyslexia, and hallucinations.

Despite the fact that the majority of ischemic strokes are a result of embolic
mechanisms, there is a complex relationship between these mechanisms and stenosis
or occlusion of the ICA, focusing specifically on ischemic regions known as *watershed
infarcts* [107]. Watershed infarcts involve the junctions between the distal fields of
arterial supply systems, where generally the *cortical* watershed regions are located
between the cortical territories of the ACA, MCA, and PCA and the *internal* watershed regions are situated between the white matter above the lateral ventricle and
the deep and the superficial arterial systems of the MCA, and between the superficial systems of the MCA and ACA. Hypoperfusion and embolism often coexist and
their pathophysiological features are interactive [26]. There is experimental evidence
that associates watershed infarction with microemboli, arising from unstable carotid
plaques or from the stump of an occluded ICA, which can travel preferentially to
watershed areas due to their distinctly small size [125]. Furthermore, there is evidence that the absence of collateral flow via the ACoA or PCoA's associated with
an incomplete or dysfunctional circle of Willis imposes an additional predisposition
to watershed infarction. Arterial lumenal narrowing and endothelial abnormalities
stimulate clot formation and subsequent embolization. Reduced cerebral perfusion
limits the ability of the bloodstream to clear emboli and microemboli and reduces
available blood flow to regions rendered ischemic by emboli that block supply arteries. The watershed regions are favored destinations for microemboli that are not
cleared.

The physiological response of the brain to reduced cerebral perfusion pressure distal to ICA occlusion has been established via physiological imaging techniques such as positron emission tomography (PET) and Single Photon Emission Spectroscopy (SPECT) [19, 50, 126]. The initial response to a decline in the CPP is an autoregulatory vasodilatation of the resistance arteries, known as stage one hemodynamic impairment [40]. This results in increased cerebral blood volume and impaired response to vasodilatory stimulation (via hypercapnia or intravenously administered acetazolamide) [19, 50, 126]. With further reduction in CPP, the autoregulatory vasodilatation becomes inadequate and the CBF decreases. As the cerebral tissue usually maintains its metabolic rate, the tissue oxygen tension decreases and OEF increases, known as *misery perfusion* or stage two hemodynamic impairment [19, 40, 42]. Below the CBF ischemic penumbra threshold, cerebral tissue function is impaired and the affected tissue is at risk of infarction [18], although it is unknown if persisting reductions in CBF above the penumbra threshold may also result in infarction.

The focus of the present study can now be placed in the context of this neurological dysfunction, as the long term goal is to devise a clinical diagnostic tool to help aid in its prevention. It may be apparent that the category of hemorrhagic stroke involves the rupture of the arterial wall, which is a complex function of the mechanics and physiological processes occurring in the arterial wall, as well as a fluid mechanics problem. Since the focus at present is the maintenance of cerebral perfusion, hemorrhagic stroke is outside the scope of the present study. Ischemic stroke frequently involves a blockage in the arteries distal to the circle of Willis (e.g the MCA in the case of large artery atherosclerosis, or the lenticular striate arteries in the case of small artery occlusion), and in most cases is the result of an embolus travelling through the bloodstream to a particular location. While it would be possible to model the trajectory of an embolus in the bloodstream and potentially predict the cerebral artery in which it may lodge, it would be extremely difficult to envisage a computer modelling technique which could predict when this may occur or what size the embolus may be. The clinical context of the present research hence lies in the prediction of the effects of large vessel atherosclerosis in the afferent arteries, proximal to the circle of Willis such, that there is the potential to provide collateral flow to perfuse all of the cerebral territories. As was outlined in Chapter 1, the particular anatomical configuration of the CoW may limit the amount of collateral flow. It is generally known clinically that certain individuals may have a complete occlusion of an ICA with no cause of neurologic impairments, while others cannot survive such a large unilateral reduction in blood supply and would instead suffer

a large ischemic stroke. With reference to the studies in Chapter 1, this ability is
most likely a function of the individual's circle of Willis.

Chapter 4

Literature Review

4.1 Introduction

The previous three chapters presented an introduction to the anatomy and physiology relevant to the modelling of cerebral hemodynamics, as well as its importance in terms of ischaemic stroke. Chapter 1 outlined the cerebral vasculature, with a description of the variations which are observed in the general population. Chapter 2 highlighted some of the important chemistry in terms of oxygen delivery to the tissues, the various control systems that the body has to regulate blood pressure and peripheral resistance, and ended with an in depth look at the physiology behind cerebral autoregulation. The purpose of this chapter is to present a review of previous work which has been done, encompassing the entire field of cerebral hemodynamic modelling.

The major difficulty in reviewing previous works lies in classifying the various aspects of each model. As a very broad generalization, models can be categorized based on whether they treat the arteries comprising the cerebral vasculature individually, or rather treat them as one lumped parameter. As an example to illustrate, the end result with the former approach is that flowrates can be calculated in the various arteries comprising the circle of Willis (and possibly distal or proximal to the circle of Willis), whereas with the latter approach the result tends to be one overall flowrate in the brain. As a second very broad generalization, models can be categorized based on whether or not they are straight fluid dynamics problems, or if they integrate mathematical models of the body's physiological controls systems (primarily cerebral autoregulation) to couple in with the fluid dynamics. As a general trend, the work incorporating the more complex physiological autoregulation models has tended to be coupled in with lumped parameter models of the cerebral vasculature, and similarly the work incorporating more complex mathematical models of the arteries in terms of fluid dynamic equations and arterial wall motion has tended to either implement very simple models of autoregulation, or none at all.

There is of course some overlap between modelling the anatomy and the physiology, and to gain a complete understanding of previous models, the coupling between both aspects needs to be considered, when it is present. Nevertheless, this generalization provides the most logical way to highlight the previous work, and this review will therefore be divided into two sections. The first will review the previous work modelling the cerebral vasculature, and the numerous aspects associated with it, while the second will review previous work modelling the autoregulation physiology, with the numerous levels of complexity associated with that.

4.2   Circle of Willis Modelling

Considering now the modelling of only the cerebral vasculature, the most logical breakdown is to categorize models based on how the governing fluid mechanics equations are formulated. Blood flow is governed by two equations physically representing the conservation of mass and the conservation of momentum (see Appendix C) and in their most general form blood flow is a function of three spatial dimensions and the temporal dimension. Depending upon what assumptions are applied to these equations at the stage of their formulation regarding the nature of the flow however, can give rise to models which are 1D, 2D, 3D and can be either steady or unsteady, with rigid or distensible walls, among other features. There is then the question of what simulations were performed with the model (i.e. what questions were the modelers trying to answer). While the nature of the simulations will be mentioned in this review, their results and significance will not be addressed (except for some specific cases where the simulations are particularly relevant to the present study), rather just the modelling aspects used.

4.2.1   1D Models

Perhaps one of the simplest forms of the governing equations arises when blood flow is described as fully developed, steady, axisymmetric flow in a straight rigid pipe, to which the continuity and momentum equations reduce to those for Poiseuille flow [169]. Using this result the network of arteries comprising the circle of Willis can be analyzed using an electrical network analogy (Figure 4.1) with a linear system of equations resulting. Some of the earlier work of Hudetz et al [67] used this analogy to create their generic 1D model of the circle of Willis (Figure 4.1) including also the anterolateral central arteries as well as lumped parameter cortical arteries and cortical anastomoses (although little information is given regarding the assignment of the resistances of these anastomoses). The network used the specification of arterial pressure at the nodal boundaries and a *Krogh cylinder* model [101] was used to determine the oxygen distribution at various locations in the arterial network. A cerebral autoregulation model included in that resistance of the cortical arteries was allowed to alter as a function of time in response to a change in flow from its reference value, although no physiology behind the autoregulatory process was included. While the Poiseuille flow relation is derived for steady state flow, it is frequently used with time varying resistances as a quasi-steady system. The simulations performed investigated reductions in afferent blood pressure as well as occlusion of an MCA on efferent blood flow.

In later work performed by Hillen et al [61], arterial blood flow was modelled as 1D unsteady pulsatile flow in a straight elastic walled tube, resulting in a set of *hyperbolic* equations. While this initial model was only applied to a single PCoA and its immediate neighbours to investigate the effect on changes in its diameter, subsequent work [62] extended the model to include the entire circle of Willis and investigated the effects of alterations in arterial diameter of the PCoA's and VA's. Only a single generic circle of Willis model was used and no autoregulation model was implemented. Using similar governing equations the work performed by Zagzoule et al [168] modelled the circle of Willis as 1D unsteady pulsatile flow in a straight elastic walled tube. Their generic model however connected all of the efferent arteries of the circle of Willis to a single lumped parameter microcirculation bed, representing the arterioles, capillaries and venules, followed by the cerebral veins (Figure 4.1). In the simulations performed the diameters of the cortical arteries were altered, as would be done through the autoregulation mechanism, however no autoregulation model was included. The results of the study were the effect of different cortical

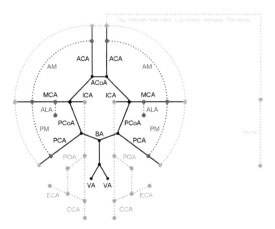

Figure 4.1: A schematic of a 1D network model of the circle of Willis. Black represents the main arteries of the circle of Willis (incorporated in all works). Purple illustrates the anterolateral central arteries (ALA) and the anteromedial and posteromedial anastomoses (AM and PM) included in the work of Hudetz et al [67] (note: the VA's were not included in this work). Pink illustrates the common and external carotid arteries (CCA and ECA) and the periorbital anastomosis included in the work of Viedma et al [164]. Green illustrates a simplification of the arterioles, capillaries, venules and veins included in the work of Zagzoule et al [168].

diameters on blood flow and pressure at various points in the vascular network.

   Almost 10 years later, the work of Cassot et al [27] again treated the flow in the circle of Willis as 1D unsteady flow in sections of straight elastic walled tubes, although in this case the pulsatile nature of blood flow was neglected. In this generic model only the arteries of the circle of Willis were included, with the specification of pressure at the boundary nodal points and no autoregulation model was implemented. In the simulations performed the effects of a combination of degrees of ICA stenosis and ACoA diameter on arterial pressures around the circle of Willis were investigated. In later work [28] the flow in the circle of Willis was simplified back to a Poiseuille flow relation in order to provide a linear system of equations, although the model was still limited to the arteries of the circle of Willis and furthermore neglected cerebral autoregulation. In a similar manner to the earlier work, the blood pressures around the circle of Willis were the primary result of the study, however the simulations were also extended to include variations in the PCoA diameter as well as ACoA diameter in combination with various degrees of ICA stenosis. As an extension of the earlier work of Cassot et al, the work of Viedma et al [164] treated the flow in the circle of Willis as 1D unsteady pulsatile flow in a straight elastic walled tube. Their model was a generic model, but differed in that in addition to the major arteries comprising the circle of Willis, the model included the carotid bifurcation with the CCA, ECA and an anastomotic connection with the periorbital artery and the ICA (Figure 4.1). Furthermore, doppler velocimetry was used to measure the velocities and subsequently specify flowrates on the afferent boundaries. The simulations performed investigated the ability of the circle of Willis and the periorbital anastomosis to provide collateral flow in response to an occlusion of the ICA. The effects of variations in communicating artery diameters on certain afferent and efferent flowrates following the occlusion were tested, although no autoregulation model was included in the model.

In the work of Ursino et al [91] a generic circle of Willis model was created using a Poiseuille flow relation and hence an electrical network analogy. The study also included lumped parameter models of the arteriole and capillary beds branching off each efferent artery, which were subsequently connected via a venous return system and also included the effects of CSF formation and outflow respectively. Perhaps the major distinguishing feature of this modelling technique is the extension of the electrical network analogy to utilize capacitors to simulate the effects of arterial distensibility. It should be understood however that this approach differs from the formulation of the governing fluid equations to encompass distensible walls. The simulations were unsteady, although non-pulsatile, and so the network analogy extends to using the impedances of various vessels rather than just the resistance (as with just using Poiseuille flow) and hence a *differential-algebraic* system of equations [57] which needs to be solved. The work included a simple autoregulation model, where the capacitance of a given efferent territory is altered in response to the flowrate deviating from its reference value. The simulations performed investigated the various degrees of cerebral vasospasm (a sudden constriction of a blood vessel) in the MCA on pressure and flow throughout the circle of Willis.

In later work [160] this model was applied to the investigation of carbon dioxide vasoreactivity (the change in arterial diameter in response to changes in concentration of carbon dioxide in the blood). While the form of the fluid equations remained the same, the effect was modelled by including a second separate mechanism in addition to the autoregulation mechanism which functioned in parallel with it, altering the capacitance of a vessel when blood carbon dioxide levels deviate from their reference level. In the study, thirty four patients (twenty healthy subjects and fourteen with unilateral ICA occlusion) were subjected to periods of hypercapnia of approximately 150s duration, while the blood velocity in both MCA's were monitored using *Transcranial Doppler Ultrasonography (TCD)*. The computer simulations performed utilized generic values for the dimensions of the arterial segments comprising the models and simulated the experimental increases in carbon dioxide. The major result from the experiments was that individuals with a unilateral ICA occlusion showed significantly lower vasoreactivity in the ipsilateral hemisphere (indicated by the increase in MCA velocity) compared to the healthy individuals who showed similar responses in both hemispheres. The computer simulations reflected this result, closely replicating the cardiac averaged MCA velocities over the course of the experiments. In addition, the computer simulations predict that carbon dioxide vasoreactivity in the ipsilateral hemisphere of an individual with a unilateral ICA occlusion will be reduced dramatically as the diameter of their communicating arteries reduces, although there was no experimental data collected regarding the individuals circle of Willis geometries to confirm this.

The later work of Charbel et al [32] illustrates one of the first attempts at making models of the circle of Willis patient specific. In a similar manner to much of the work mentioned previously, the blood flow is treated as 1D unsteady pulsatile flow in sections of straight elastic walled tube. The model differs from previous work however, in that in addition to including the arteries of the circle of Willis, the model extends proximally back to the aortic arch. The model uses *digital subtraction angiography* (see Chapter 5) to obtain the diameters of the cerebral vessels and phase contrast MRI (see Appendix A) to determine the reference blood flowrates through the various afferent and efferent arteries, utilized as a calibration for the model. The model was applied to sixteen patients, predicting the ability of their circles of Willis to provide collateral flow during a *balloon occlusion test (BOT)* (the temporary procedure where an ICA is occluded with an inflatable balloon and collateral circulation through the circle of Willis is assessed to determine whether or not the occlusion can be made permanent in order to reduce the risk of aneurysm rupture), where the model results were the efferent flowrates following the ICA occlusion, and a reduction in ipsilateral CBF of more than 25% was considered a failure of the test. Despite the quantitative nature of the computational simulations of the BOT, the experimental assessment of the BOT was made based on a qualitative inspection of the patients neurological state, monitoring features such as cranial nerve function, muscle strength, and language ability. Thus, although the results of the study showed 100% agreement between numerical and experimental pass or failure of the test (ten patients passing the test and six patients failing), there is no data to gauge how closely the computer model actually predicted the efferent flowrates following the BOT. Furthermore, although the autoregulation mechanism would play an important role in restoring blood flow following an ICA occlusion, no autoregulation mechanism was implemented in the model.

In the work of Alastruey et al [10], cerebral blood flow was 1D unsteady pulsatile flow in sections of distensible tube. This model differed from others in that it included the aortic arch, subclavian and common carotid arteries. Furthermore, since arterial wave propagation was of interest, a 0D electrical circuit analogy known as a Windkessel model, was incorporated as a boundary condition at the efferent terminations because it created a non reflective coupling between the 0D and 1D segments. The simulations performed examined a number of circle of Willis configurations, occluding an ICA and a VA and investigating the subsequent effect on cerebral flowrates. In later work [9] the model then incorporated the autoregulation mechanism developed by Moore et al [109], which was used to simulate the response to a unilateral pressure drop in an ICA.

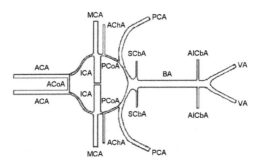

Figure 4.2: A schematic of the 2D model of the circle of Willis employed in the work of David et al [36, 45].

4.2.2   2D Models

With slightly fewer assumptions applied to the conservation of mass and momentum equations they can be derived in a 2D unsteady formulation. The transition from 1D to higher dimensions introduces the added complexity of designing a surface model representing the structure of the circle of Willis, which must then be discretized into a computational grid. It should be understood that while many of the 1D models presented involved the discretization of the 1D arterial segments comprising the cerebral vasculature modelled, it is a much more trivial task to divide 1D lines into segments compared to the complicated geometry of the circle of Willis. In the work of Ferandez et al [36, 45] this approach was used where the circle of Willis was treated as an unsteady non-pulsatile 2D rigid walled structure. The generic model included the main arteries of the circle of Willis, with the addition of the AICbA's, SCbA's and AChA's (Figure 4.2) and used the novel approach of placing a lumped parameter 'porous block' at the efferent artery terminations, which constituted an extra body force term in the momentum equation. The actual boundary conditions used were the specification of blood pressure at the afferent inlets and the distal end of the porous blocks. The model implemented a simple autoregulation model where the permeability of the porous block (simulating peripheral resistance) was altered when blood flow deviated from its reference level. A number of configurations of the circle of Willis were examined by removing arterial segments from the generic structure and simulating the transient response of blood flux (as actual flowrate is not as well defined in 2D) in the efferent arteries to a step reduction in afferent blood pressure.

4.2.3   3D Models

Using the governing equations in their most general form requires the most complicated methods of model and computational grid generation, but does provide the most complete description of the hemodynamics, and furthermore, lends itself to the use of medical imaging data to create the 3D models. The work of Cebral et al [29] illustrates one of the first attempts at the creation of patient specific models of hemodynamics, treating the flow as unsteady pulsatile flow in a 3D rigid walled model. The model was generated from Time of Flight MRI data, including only the major arteries of the circle of Willis, and a finite element scheme was applied to the subsequent computational grids. The boundary conditions used phase contrast MRI to measure flowrates on the afferent arteries which were subsequently transformed into a velocity profiles by the superposition of *Womersley profiles* [169]. On the efferent terminations a 1D electrical network analogy was employed where an overall peripheral resistance could be computed assuming the sum of an infinite geometric series of branching resistors and from this a pressure could be determined to specify on the efferent boundary. The work was aimed at predicting the efferent flowrates through the circle of Willis in response to a temporary occlusion of the A1 segment of an ACA, as would be performed during a surgical procedure to clip an intracranial aneurysm. Despite the major result of the simulation being how blood flow to the various territories is affected by the occlusion, no autoregulation mechanism was incorporated and the study only involved a single test on one circle of Willis model.

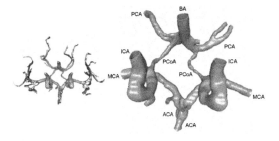

Figure 4.3: The MRI segmented 3D circle of model used in the work of Cebral et al [31].

In later work [30] the modelling technique was extended to allow for distensible arterial walls and while it was only tested on simple arterial segments such as the carotid bifurcation and the renal arteries, it was later applied to model the circle of Willis [31]. In this study the same methodology was used and the model consisted of only the major arteries of the circle of Willis (Figure 4.3). This later work did however use the novel approach of 'growing' 1D fractal networks at the termination of the efferent arteries of the circle of Willis in order to determine a peripheral resistance to the blood flow. The approach involved determining a cerebral tissue volume considered to be perfused by a given efferent artery via the MRI data and using a constrained constructive optimization technique [76] to grow the fractal tree in that volume. The afferent boundary conditions were then replaced with the superimposed Womersley profiles. The model did not include any autoregulation mechanism and only one generic circle of Willis model was investigated in the study, the reason being that it was the modelling methodology which was the focus of the paper rather than its application to investigate any clinical related matters.

In the later work of Kim et al [80] the blood flow was treated as 3D unsteady

Figure 4.4: The MRI segmented 3D circle of Willis model used in the work of Kim et al [80].

pulsatile flow in a distensible walled model. The model was generated from MRI Time of Flight data (Figure 4.4) and used the specification of pressure at the afferent boundaries and included a 1D electrical network analogy at the efferent terminations where an overall peripheral resistance could be computed assuming the sum of an infinite geometric series of branching resistors. The model also included a simple autoregulation mechanism which altered the overall peripheral resistance as the flowrate through a given efferent artery deviated from its reference level. Simulations were performed only on one circle of Willis model investigating the transient efferent flowrates in response to a step reduction in afferent blood pressure in an ICA.

4.3   Cerebral Autoregulation Modelling

Due to its clinical importance in terms of cerebral hemodynamics, there has been a significant amount of work performed developing mathematical models of cerebral autoregulation to aid in general understanding and to be used as clinical tools. The way in which the models are constructed is even more variable than modelling of the circle of Willis, but there are two broad generalizations that can be made to categorize them. The first is to consider models which are based on attempting to describe the underlying physiological factors driving autoregulation. As was mentioned in Chapter 2, the complete chemical pathways leading to the alterations in smooth muscle tone (and hence cerebrovascular resistance) and their interaction are not completely elucidated and so these types of models tend to include a mathematical description of the physiology up to a certain point and then incorporate phenomenological lumped parameter variables to complete the link between a catalyst for autoregulation and an alteration of resistance. These physiology based models also tend to include a simple lumped parameter resistance representing the cerebral vasculature as a whole, but this simplification leads to the ability to include other aspects of cerebral mechanics such as the effect of increased blood vessel diameter on cerebral blood volume and hence intracranial pressure.

The other broad generalization is to categorize models that aim instead to essentially develop transfer functions or neural network models which fit experimental measurements made regarding autoregulation variables such as continuous measures of arterial blood pressure and cerebral blood flow velocity for example. While not necessarily deepening the understanding of how autoregulation works, or allowing for certain types of physiological simulations to be performed, this approach can often give useful information about the frequency response of autoregulation and has the advantage of using experimental autoregulation data from humans which can often reproduce the brain's physiological responses reasonably well. As it is the aim of the present study to use the former approach for implementing an autoregulation model, this section will give a review of other autoregulation models which have been developed with the physiological processes in mind. Autoregulation models using the latter approach will not be examined other than mentioning that a significant body of work has been performed by Panerai et al [117, 118, 119, 120, 121], Olufsen et al [114, 116] and Mitsis et al [102, 103, 104, 105, 106] generating and validating these types of models.

One of the most significant contributions to the field of cerebral autoregulation modelling has been made by Ursino et al, producing a variety of models of varying complexity and applications. One of the earlier works [156] began by treating the cerebral tissue as parallel plane layers and developing partial differential equations to describe the diffusion of oxygen from the capillary blood into the cerebral tissue. The work then continued by making an assumption about a 2D capillary network within a layer (Figure 4.3) and deriving partial differential equations for the diffusion of vasoactive metabolites from the cerebral tissue into the perivascular space due to a lack of oxygen. By then making assumptions regarding the nature of the equations, the partial differential equations were simplified down to ordinary differential equations in time. The equations for the diffusion of metabolites were then applied specifically to H^+ and adenosine and as a starting point, the simulations were performed to specifically examine the time dynamics of the tissue pH, tissue oxygen tension and adenosine concentration in response to a changes in CBF. In the second part of the work [157] these ordinary differential equations for pH and adenosine were combined with a model of the cerebral vasculature, treating it as five resistors in series, representing five distinct levels of the vascular bed (Figure 4.5(b)). The model incorporated a Poiseuille relation between pressure, resistance and flowrate, Laplace's law [1] between pressure, arterial radius and wall tension and a phenomenological relation between smooth muscle tension and pH and adenosine concentrations, thereby assuming that it was these metabolites driving autoregulation. The simulations performed examined the temporal changes of pH, P_tO_2 and adenosine at distinct vascular levels in response to changes in arterial and venous pressure, arterial oxygen concentration, and reactive hyperemia.

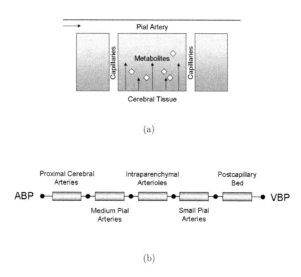

Figure 4.5: (a) The 2D model of the cerebral tissue and blood vessels used in the work of Ursino et al [156] and (b) the division of the cerebral vasculature used in the subsequent work [157].

In a subsequent extension of this model [153] a myogenic and neurogenic mechanism were included (Figure 4.6) which implemented phenomenological relations affecting the overall smooth muscle tension in addition to the pH and adenosine equations. The myogenic relation treated its response as a function of the stress in the blood vessel walls and the neural relation as a function of arterial and intracranial pressure, essentially simulating part of the baroreceptor mechanism. Since the neural mechanism was a new addition the first block of simulation results was aimed at examining the effects of artificially 'cutting' the neural feedback loops on arterial radius and blood flow. The simulations then investigated the effect of changes in arterial blood pressure on CBF, producing autoregulation curves at various allowable times for the overall autoregulation mechanism to respond to the change in blood pressure.

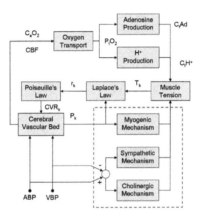

Figure 4.6: A schematic of the autoregulation model implemented in the work of Ursino et al [153], illustrating the various components of the autoregulation mechanism and its relationship between blood flow, pressure and smooth muscle tone.

In other work [151] a 1D electrical network analogy was used to model the cerebral vasculature, breaking it up into four distinct levels, where Poiseuille's law was used to relate pressure, flow and resistance, as was done previously. The model differs, in the use of capacitors and diodes to replicate blood vessel compliance and the one way flows of blood across the blood brain barrier into the CSF and back into the venous system respectively. The autoregulation model was a simplification of previous works considering the change in resistance to be a function of simply a deviation in blood pressure from a reference value. While this work was the first part of a dual publication, outlining the model development, simulations were performed to simulate the genesis of intracranial pulse pressure from the pulsatile arterial blood pressure, investigating the relationship between pressure waveforms at various points in the cerebral vasculature and at different levels of intracranial pressure. The second part of the work [152] involved performing a number of simulations investigating the effect of certain clinical conditions, such as injections of fluid into the CSF on blood flow and offered some validation with experimental work. As a later extension of this model [155] the 1D electrical network analogy was modified by further subdividing the vasculature into five distinct levels (Figure 4.3) and including Laplace's law in combination with Poiseuille's law to relate pressure, flow, resistance, arterial diameter and wall tension. The autoregulation model was modified from the previous model, but was still a simplification compared to earlier works [153, 156, 157] by considering the physiological mechanism to be made up of two parts. The first is essentially a myogenic mechanism where the drive for changes in wall tension occurs when blood pressure at the different levels differ from a reference level. The second part is essentially a simplification of the relation between blood flow and the formation of metabolites where the drive for changes in wall tension occurs when blood flow deviates from a reference level. Part of the work involved examining the generation of *plateau waves* (a sudden rapid elevation and subsequent reduction of ICP brought about by a closed loop unstable process linking active vasodilation, increased cerebral blood volume, elevated ICP and decreased cerebral perfusion pressure), while the remainder of the simulations was aimed at investigating autoregulation itself and so part of the simulations involved examining the effects of a number of step reductions in arterial blood pressure on CBF.

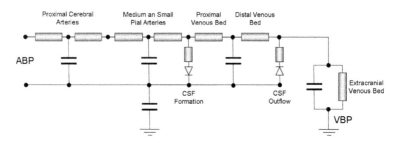

Figure 4.7: A schematic illustrating the electrical network analogy implemented in the work of Ursino et al [155].

Later work involved variations of the electrical network analogy using slightly different distinctions between levels of the vasculature and different implementations of the cerebral autoregulation mechanism. In another body of work [159] autoregulation was taken to be a function of a deviation in CBF from a reference value, similar

to other models [155] but different in that this driving force for changes in vascular
tone was implemented by changing the arterial compliance rather than resistance.
Then, in later work [90] another 1D electrical network model was proposed, again
combining Poiseuille's and Laplace's laws. The extension in this work was that in
addition to incorporating an autoregulation mechanism which was a function of a
deviation in CBF from a reference value, a second mechanism was introduced to
account for carbon dioxide vasoreactivity by making the drive for autoregulation
a function of a deviation in arterial carbon dioxide concentration from a reference
value. Both of these mechanisms sought to alter arterial wall tension, and the major
result of the simulations was to examine the temporal effects of variations in arterial blood pressure and carbon dioxide on CBF and ICP. This modelling technique
was then tested in later work [163] against experimental data in humans, regarding
changes in arterial blood pressure carbon dioxide with measurements of ICA velocity
to correlate to CBF.

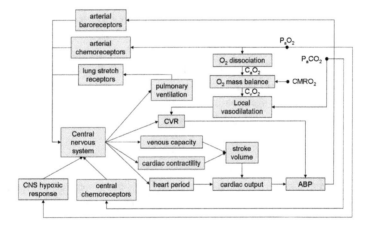

Figure 4.8: A schematic illustrating the cardiovascular model with additional physiological control systems and their relationships, implemented in work of Ursino et al [96].

In subsequent work [161, 162] the 1D electrical network analogy has been extended to the entire cardiovascular system, including the pulmonary circulation and
a subdivision of the systemic circulation to the heart, brain, skeletal muscle, viscera
and extraviscera. The model hence included models of the heart and pulmonary
ventilation as well as a number of physiological control systems in addition to autoregulation, such as the arterial baroreceptor and chemoreceptor loops and the

CNS ischemic response. The autoregulation model in this work differed from pre-vious models however in that it assumed that the driving force for autoregulation was due to a deviation in the venous oxygen concentration from a normal reference value. The venous oxygen concentration was calculated based on the arterial oxygen concentration, an overall metabolic rate for the brain, and the CBF, and elicited an autoregulatory response via change in resistance. While the first part of the two part publication [161] was aimed at the model development, some sample simula-tions were performed illustrating the transient response of flowrates, pressures and heart rate to deep hypoxia. The second publication [162] then sought to validate the model predictions against experimental data, mainly examining the steady state re-sponse of flowrates throughout the model and cardiac output over a range of oxygen tensions. The results of the computational simulations showed a close agreement with experimental data (taken from other clinical studies using human subjects) in terms of variables such as HR, CO, ABP, and PR following periods of hypoxia.

In other work [96], the same basic model of the cardiovascular system was in-corporated (Figure 4.3); the major point of interest being that the autoregulation model modified CVR based on a deviation of both venous oxygen concentration and arterial carbon dioxide tensions from their normal reference values. This model was used to investigate the effects of hypocapnic hypoxia in both healthy and anaemic subjects, producing steady state variations in cardiac output, arterial pressures, and flowrates throughout the model over a range of oxygen tensions.

In the work of Lu et al [94] the 1D electrical network analogy adopted from the work of Ursino et al [163] was implemented to describe the cerebral vasculature, but incorporated partial differential equations in the blood vessel compartments describing the diffusion of oxygen and carbon dioxide into the cerebral tissue and CSF. The autoregulation model assumed that the changes in arterial compliance (the autoregulatory response) were based simply on a deviation in CBF from its normal reference value, but differed from other works in that the arterial carbon dioxide tension altered this reference CBF such that the effects of carbon dioxide vasoreactivity could be incorporated. The model also included a chemoreceptor loop, altering the depth and frequency of the *pleural* pressure variations in response to changes in pH in the CSF. The model was used in a wide range of simulations mainly examining the transient responses of flowrates and pressures (among other parameters) to maneuvers such as hypotension, a thigh cuff test and carotid artery compression.

In the work of Thoman et al [144, 145] models of the intracranial dynamics have been created with the aim of developing 'patient simulator' software for use as a clinical and learning tool. Unlike the electrical network analogy frequently adopted by Ursino et al which distinguished between various regions of the cerebral vasculature, this model treated the cerebral vasculature as a single lumped parameter resistance to flow, incorporating simple relationships between various clinical variables (Figure 4.9). The model incorporated a Poiseuille flow-type relation between pressure, flowrate and resistance, although it is not strictly Poiseuille flow as resistance is used in a broader sense (a result of the lumped parameter resistance) and not related to any arterial diameter in any way. Furthermore, rather than using capacitors to simulate the distensibility of the cerebral arteries, a simple empirical relation was used to relate CBV and ICP, with a feedback loop to affect CBF. The relevant part of the model is the autoregulation mechanism, which defined the driving cause for alterations in CVR to arise from a deviation in CBF from its normal reference value. While this is a common approach, this model differs in that the reference level of blood flow is a function of the amount of oxygen and carbon dioxide in the blood, as well as the brain's metabolic rate and the oxygen extraction fraction (the amount of available oxygen actually being used by the cerebral tissue). The simulations performed were essentially aimed at validating the model and involved comparing its predictions of CBF at various concentrations of arterial carbon dioxide and MABP to experimental measurements.

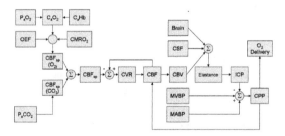

Figure 4.9: A schematic of the cerebral vascular compartments and relationships implemented in the patient simulator autoregulation model of Thoman et al [144].

The work of Jung et al [73] treated the cerebral vasculature in a similar manner to Ursino et al [159], using a network analogy to describe the four distinctions made regarding the blood vessel levels in terms of resistances and capacitances. The model also included the brain tissue and CSF as separate compartments in the model so that the effects of changes in CBV on CSF production and ICP could be investigated. The autoregulation mechanism implemented in the study altered the capacitance of the blood vessels, assuming that the driving force for autoregulation was a deviation in CBF from its normal reference value. As an extension to the work of Ursino et al however, the model also incorporated a Krogh cylinder model [101] to determine the oxygen tension within the cerebral tissue based on the oxygen concentration in arterial blood. The simulations performed examined CBF and CSF production as well as tissue oxygen tension over a range of arterial blood pressures, both with and without the autoregulation mechanism active to investigate its effects on these variables.

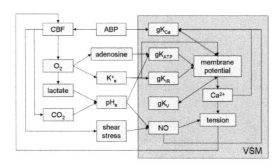

Figure 4.10: A schematic of the physiological processes involved in the autoregulatory response in the work of Banaji et al [16].

Perhaps one of the most complete mathematical descriptions of the physiology behind the autoregulatory process has been the work of Banaji et al [16]. In their model, the nature of the cerebral vasculature was based on the 1D electrical network analogy developed by Ursino et al [90] but modified the link between the blood flow, pressure and the development of tension with much more detailed physiological processes occurring. The model assumed that the initial driving force for changes in smooth muscle tension (Figure 4.10) was blood flow, through changes in the concentration of oxygen and carbon dioxide (as well as affecting shear stress on the endothelial wall), leading to the production of certain vasoactive metabolites, which affect the activation of various potassium channels on the smooth muscle cell plasma membrane, ultimately leading to changes in intracellular calcium concentration, force development by the actin-myosin myofilaments, and hence vascular tone.

The model utilized five compartments, dividing the brain into the vascular lumen, vascular smooth muscle, extracellular space, intracellular space and the mitochondrial matrix (Figure 4.11), describing the chemical reactions relevant to that compartment by either conservations equations where the physiology and biochemistry is understood, and phenomenological relations for any remaining reactions. The vascular lumen compartment included a series of reactions of hemoglobin with oxygen, carbon dioxide and hydrogen ions as well as other reactions involving lactate. The vascular smooth muscle compartment included a number of phenomenological relations between the concentration of certain metabolites and the activation of ion channels in the plasma membrane, as well as phenomenological equations for intracellular calcium concentration based on the activation of the ion channels, phosphorylation of the myosin light chains and the subsequent overall 'activation' of the muscle coupled with the 1D model. The remaining three compartments, comprising the brain tissue, involved a number of conservation equations describing the various biochemical pathways involved in metabolism such as glycolysis, ATP phosphorylation, the TCA cycle and the electron transport chain.

Since the model itself made such a large and complex extension to the mathematical description of the autoregulatory process, the work was devoted mainly to the model development. Some sample simulations were performed however, examining the steady state CBF in response to changes in ABP and arterial carbon dioxide tension, as well as comparing model predictions for *tissue oxygenation index (TOI)* following a head-up tilt manoeuver to experimental data.

The conclusions which can be drawn from this review are that to date, there has been a reasonable body of work performed on modelling blood flow in the cerebral vasculature (although the majority are 1D models), as well as modelling the physiological control systems utilized by the body to regulate cerebral blood flow. However, there has been very little research aimed at integrating both the complex 3D cerebral hemodynamic models with an autoregulation model, leaving a large scope for this type of modelling to 'bridge' the two fields. Furthermore, there has been very little work investigating the effect of the different circle of Willis configurations on cerebral blood flow, and it will be the goal of the present study to cover both of these aspects of cerebral hemodynamic modelling.

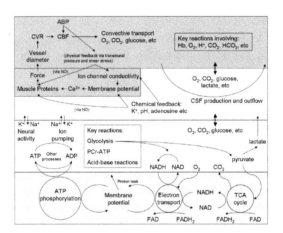

Figure 4.11: A schematic of the five compartments and the associated chemical pathways implemented in the autoregulation model of Banaji et al [16]. Orange represents the vascular lumen compartment and the associated blood biochemistry, pink represents the vascular smooth muscle compartment and the reactions involving actin and myosin myofilaments in the production of force, light blue, yellow and green represent the extracellular, intracellular and mitochondrial parts of the cerebral tissue respectively, involving a series of reactions related to metabolism as well as the movement of ions and other metabolites.

Chapter 5

Geometry Generation

5.1 Introduction

The previous four chapters provided an introduction to the anatomy, physiology and pathology relevant to cerebral hemodynamics, as well as a review of previous works in the field. With this information outlined, the next three chapters will cover the details of the model development, beginning with the generation of the 3D circle of Willis geometry. To aid in the justification for certain modelling decisions later on, it is worth clearly stating the goals and purpose of the present study. These are to:

- Develop a 3D computer model to investigate the effect of arterial geometrical variability of the circle of Willis on cerebral blood flow

- Develop a technique by which patient specific models of the circle of Willis may be generated and used to simulate the response to clinical scenarios, such as a reduction in afferent blood pressure from stenosis or occlusion of an artery. The technique should also allow for extension in future work to model the effects of embolic and hemorrhagic stroke

As was illustrated in Chapter 1, the cerebral vasculature is a highly complex three dimensional structure which varies among the general population. In order to simulate blood flow throughout the brain to a level of detail that will provide useful information, both for the purposes of pure research and as a clinical diagnostic tool, three important questions arise regarding the development of the computational geometrical models;

- What type of anatomical data to base the model on?

- How much of the cerebral vasculature to include in the model?

- Should the model be a generic model, based on dimensions averaged from a population study or should a series of patient specific models be created?

The second question essentially focusses on the placement of the afferent termination (the inlets to the model) and the efferent termination (the outlets of the model) in the arterial network and which arteries to include in between. As it happens, the first two issues are intimately linked, as the size and the amount of detail that can be placed into a model depends upon the anatomical data used. The two main options are to use either data from autopsy studies or medical imaging data, in which case the only two contenders are *Magnetic Resonance Imaging (MRI)* or *Digital Subtraction Angiography (DSA)*. While autopsy data can provide information regarding the lengths and diameters of the arterial segments around the circle of Willis, these are averaged from population studies, which automatically implies the generation of a *generic* model. Furthermore, this data does not provide much information regarding bifurcation angles and complex 3D paths that these arterial segments take through the body. As a result the decision was made to use medical imaging data for the present study, which, as will be discussed at a later stage is essentially the only way of creating a modelling technique which may be patient specific.

MRI uses a spatially varying magnetic field and radiofrequency electromagnetic radiation to generate the phenomenon of *nuclear magnetic resonance* within hydrogen nuclei of the various tissues present in the human body. The complex MR signal generated during resonance is converted into an induced voltage signal by the principal of Faraday induction and can be used to reconstruct an image of these tissues (see Appendix A for an detailed explanation of MRI). One particular angiographic MRI technique is known as *Time of Flight (TOF)* where the MR signal from stationary tissue is suppressed and the signal from moving tissue such as blood is enhanced. TOF scans are routinely used in imaging of the cerebral vasculature and produce scaled images of resolution and quality sufficient for clinical diagnosis (Figure 5.1(a)), but suffer from the practical limitation of the maximum resolution that can be achieved. For example, with a 1.5T MRI scanner it is difficult to achieve a resolution much finer than approximately 0.5mm while still maintaining an acceptable *signal to noise ratio (SNR)*. Between the aorta and the circle of Willis this resolution limit would not be a problem, but in the brain, this resolution limit combined with the fact that there is little flow through many of the smaller cerebral arteries (reducing the effectiveness of the TOF technique), means that many of them are completely undetectable in the scan and even the major cerebral arteries (i.e. the ACA's, MCA's and PCA's) tend to lose visibility at some point along their A2, M2 and P2 segments respectively (Figure 5.1(a)).

DSA uses a combination of X-rays and an injected contrast agent to enhance the visibility of the arteries, since the contrast agent affects the X-ray attenuation through blood vessels. DSA scans provide incredibly detailed images of the cerebral vasculature all the way down to the cortical branches on the surface of the cerebrum (Figure 5.1(b)). The technique suffers from the practical drawback however that the contrast agent can only be injected into one of the major afferent arteries at a time (such as a single ICA illustrated in Figure 5.1(b)), making the generation of a model of the entire circle of Willis difficult.

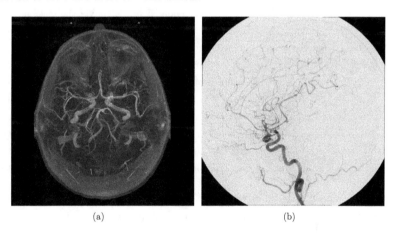

(a)   (b)

Figure 5.1: (a) A transverse *maximum intensity projection (MIP)* of the cerebral vasculature, obtained from a 3D time of flight scan and (b) a sagittal view of a DSA scan of the arteries perfused by a single ICA, showing much greater detail of the cerebral vasculature. Note: the definition of a maximum intensity projection is given in Appendix A.

In order to provide the most accurate and patient specific pathological picture of cerebral blood flow, an ideal model of the cerebral vasculature would begin at

the aortic arch, following the common carotid and vertebral arteries and include the
anastomosis at the circle of willis, continuing downstream through the branching
of the major cerebral arteries to their cortical branches. Furthermore, the model
would include all of the smaller arteries, branching off the main structure during
this course. Such a model would provide detailed information about the amount of
blood perfusing specific areas in the brain, and could also be used to predict the
response to pathological events in essentially any region of the cerebral vasculature.

In terms of investigating and modelling embolic stroke, it would be necessary
to include the smaller arterial branches in the model as emboli are frequently small
enough to only block arteries of this size. To investigate and model the effects
of thrombotic stroke (or any other cause of a reduction in afferent blood supply
to the brain) it is acceptable to model only the major cerebral arteries, placing
the efferent termination immediately downstream of the circle of Willis. Doing so
assumes that any daughter branch downstream of a major cerebral artery will suffer
the same loss of blood flow as its parent artery, as would any arteries originating
near the circle of Willis, such as the anterior choroidal or ophthalmic arteries. The
way in which this assumption may be rendered invalid is via the existence of an
anastomotic connection downstream in the *cortical* branches providing a collateral
pathway through which a significant amount of blood may be rerouted. While these
collaterals were shown to exist in Chapter 1, they all involve blood being rerouted
through a capillary network and are hence much less effective when compared to
the circle of Willis. Furthermore, it is thought that the ability to use the collaterals
depends on the perfusion pressure at that level, which is determined farther up the
arterial network at the level of the circle of Willis. As there is also no method
at the present time to quantify the significance of these cortical anastomoses it is
considered a valid modelling assumption to ignore their effects for the present study.

To represent the effects of the resistance arteries, arteriolar, venous and capillary beds, eliminated from the model by the choice of efferent artery termination,
the approach of using a *porous block*, proposed by Ferrandez et al [37, 46], at the
termination of each efferent artery has been implemented. The *resistance* of the
porous blocks can be set to provide a resistance to the flow in much the same way
as the vascular beds, hence realistic representations of the hemodynamics can still
be achieved. Considering now the larger arteries feeding into the brain, the afferent
terminations have been placed below the *carotid siphon* of the internal carotid arteries and either at the beginning of the basilar artery or at the level of the vertebral

arteries, immediately before their fusion into the basilar artery. The assumption is therefore made that any reduction in afferent blood supply through the common carotid and vertebral arteries will be propagated through to the internal carotid and basilar arteries. In keeping with the goal of investigating the effects of the variability of the circle of Willis on cerebral hemodynamics, initial work began by developing idealized models with a subsequent shift in focus to developing techniques for the creation of patient specific models. This chapter will give a detailed outline of both modelling techniques.

5.2 Idealized Geometry

The idealized geometries were created in the *computer aided design (CAD)* package *Solidworks* and the explanation of how the models were created will use the terminology adopted by Solidworks for certain geometric features. The tools used within the software to create the models are standard CAD features meaning that the technique could readily be transported to any commercial CAD software. The procedure for creating an idealized 3D circle of Willis model from an MR TOF angiogram began with spatially aligning three orthographic maximum intensity projection images (Figure 5.2(a)), which were scaled to the *field of view (FOV)* of the scan. Afterwards, splines were traced over the afferent, efferent and communicating arteries comprising the CoW (Figure 5.2(b)). Using the splines from any two of the three projections available a 3D *projected curve* was created for each artery (or section of an artery) (Figure 5.2(c)). This procedure resulted in a 3D skeletal structure (Figure 5.2(d)) which was the generic base structure for *all* the idealized models created.

Although the splines traced over the MR projections were comprised of a series of points, the projected curves did not inherit them. Since points along the skeletal structure were a desired feature for the model construction, the projected skeletal structure was made into a *3D sketch* using the *convert entities* feature. The 3D sketch was then modified to have points placed at various intervals along the curvature of the artery (Figure 5.3(a)). By using a point and a normal vector as one possible definition of a plane, a series of planes were placed along each arterial segment at each point inserted in the 3D sketch and the direction vector of the projected curve at that point. Subsequently circular sketches were then inserted

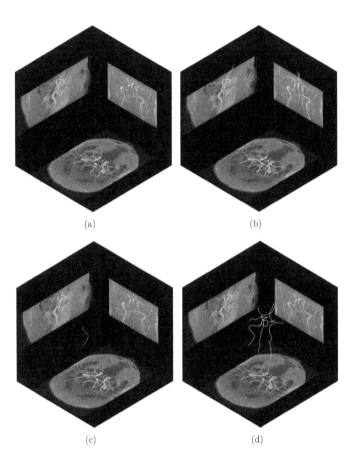

Figure 5.2: The procedure for creating the skeletal structure for the idealized models. (a) three orthogonal MR projections, (b) splines traced over the arterial segments, (c) the projection of any two orthogonal splines to create a projected curve of an artery or piece of an artery and (d) the complete skeletal structure.

on every plane with their centre located at the points along the 3D sketch (Figure 5.3(b)). An arterial segment was then completed by creating a *lofted surface* connecting the circular sketches (Figure 5.3(c)) and this process was repeated for every arterial segment in the 3D sketch.

The next step is the creation of the arterial junctions connecting the model

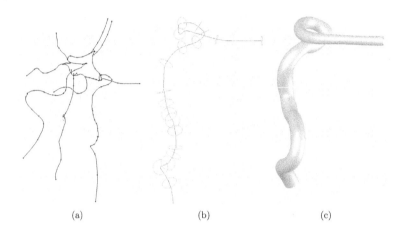

(a)   (b)   (c)

Figure 5.3: (a) The projected curve created into a 3D sketch with points then placed along the arterial segments, (b) circular cross sections sketched on planes placed along the 3D sketch and (c) the surface loft of the circular cross sections to create the arterial wall.

together. Two different methods were employed, based on the diameters and angles between the vessels forming the junction. For junctions where the two daughter arteries were of a similar size to the parent artery and had a small bifurcation angle between them (such as the VA-BA junction and the BA-PCA junction), the optimal way to create a realistic junction was to loft three crescent shaped surfaces between the three arteries comprising the junction (Figure 5.4(a)) and then use a *surface fill* (Figure 5.4(b)) to complete the junction (Figure 5.4(c)). For arterial junctions which involve a large variation in diameter between the connecting arteries and a large bifurcation angle (such as the ACA-MCA junction and all of the communicating artery junctions), the optimal method was to initially extrude a circular surface of slightly larger diameter than the smaller daughter artery up to the surface of the larger parent artery and then use the extrude to *trim* the larger artery, leaving a hole in the surface (Figure 5.5(a)). The edge of the trimmed surface was then lofted to the end of the smaller daughter artery (Figure 5.5(b)) to complete the junction (Figure 5.5(c)).

The final step in the model creation is the generation of the boundaries of the model. The afferent inlets were created by using the *planar surface* feature in order to cap the afferent terminations. While a similar procedure was adopted for the efferent terminations, these arteries then had porous blocks extruded from the

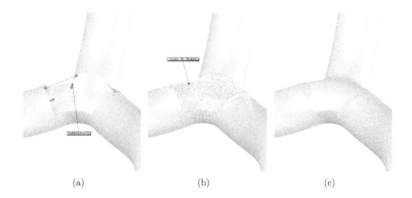

(a)   (b)   (c)

Figure 5.4: The procedure for creating junctions when the parent and daughter arteries are of similar diameter and small bifurcation angles. (a) crescent shaped surface lofts, (b) filling the surface and (c) the completed junction.

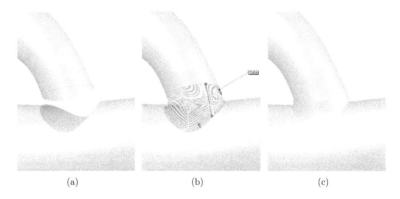

(a)   (b)   (c)

Figure 5.5: The procedure for creating junctions when the parent and daughter arteries are of greatly differing diameter and large bifurcation angles. (a) An extruded circular surface used to trim the parent artery, (b) the edge of the hole lofted to the end of the smaller daughter artery and (c) the completed junction.

planar end caps, thereby completing the model (Figure 5.6).

Generating the geometry using this technique, as opposed to direct segmentation and reconstruction from MRI data produces a physiological approximation to the cerebral vasculature, with only the loss of the finer topological details of the surface of the arterial walls. Compared to the lengths and diameters of the arterial branches

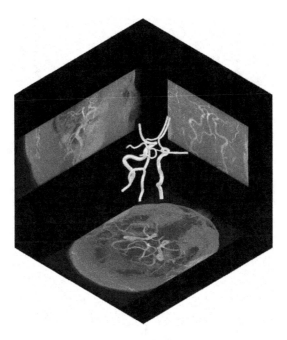

Figure 5.6: A complete idealized 3D model of a normal complete circle of Willis.

however, these topological features will have a minimal effect on cerebral blood flow. Considering both the geometrical variability and number of the arterial segments comprising the circle of Willis, investigating the sensitivity of the hemodynamics to different geometries could result in a large number of possible simulations. Three of the more common pathological conditions of the circle of Willis [13] were chosen for study, namely a complete circle of Willis, a fetal P1 and a missing A1. The difference between a Fetal P1 configuration (Figure 5.7(a)) compared to a complete circle is that one of the P1 segments of a PCA is greatly reduced in size, while the PCoA ipsilateral to that P1 is increased in size. In the case of a Missing A1 (Figure 5.7(b)), one of the A1 segments from an ACA is either completely absent from the cerebral vasculature, or is decreased to such a size that there would be negligible blood flow through it. Furthermore, in this condition there is essentially no ACoA, rather the remaining A1 bifurcates into the two A2 segments.

As the flow through the blood vessels depends much more strongly on the diam-

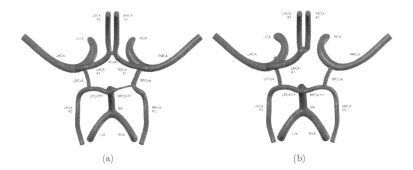

(a)   (b)

Figure 5.7: Axial views of the (a) Fetal P1 and (b) Missing A1, idealized circle of Willis anatomical variations created. Note: the models have been shown truncated part way along the internal carotid and basilar arteries to provide a clearer view of the efferent and communicating arteries.

eter as opposed to the length, it was decided to use only a single generic 3D skeletal circle of Willis structure for all of the geometries and alter only the diameters. The diameters of the various arterial segments of the circle of Willis were obtained from a retrospective population study of MRA scans. The diameters were measured using the in-house software package *Volume Estimation Toolkit (VET)* (Canterbury District Health Board), which allows various slices from a given MRI TOF scan to be read and by interactively dragging a cursor between points of interest, one can obtain measurements of the vessels. For the complete circle of Willis 13 cases were examined, for the fetal P1 pathological condition 14 cases were examined, and for the missing A1 pathological condition 10 cases were examined. The mean values for the diameters used to create each model and their standard deviations are shown in Table 5.1 where the terms ipsilateral and contralateral refer to the same side and opposite sides of the CoW as the pathological condition respectively [110].

The computational grids for the idealized models were generated in the preprocessor *Gambit* using the *TGrid* scheme to create an unstructured tetrahedral mesh (Figure 5.8). Due to the complexity of the geometry the model was separated around the junctions into arterial segments and junction segments, each of which were meshed separately. Each 3D mesh generated comprised approximately 1 million tetrahedral volumes with a nodal spacing ranging in size from 0.2mm in the smaller efferent arteries and communicating arteries to 0.35mm in the larger afferent arteries.

Table 5.1: Circle of Willis measurements used for the creation of the idealized geometries. Note: *ips* and *cont* are abbreviations for ipsilateral and contralateral.

Artery		Complete CoW		Fetal P1		MissingA1	
		Diameter	Std Dev	Diameter	Std Dev	Diameter	Std Dev
		mm	mm	mm	mm	mm	mm
ACA - A1	*ips*	2.33	0.22	2.16	0.40	−	−
	cont	2.33	0.22	2.38	0.41	2.38	0.41
ACA - A2		2.40	0.31	2.50	0.31	2.55	0.15
MCA - M1	*ips*	2.86	0.17	2.95	0.29	2.91	0.48
	cont	2.86	0.17	2.99	0.20	2.88	0.28
PCA - P1	*ips*	2.13	0.25	1.55	0.40	1.88	0.41
	cont	2.13	0.25	2.99	0.46	2.19	0.39
PCA - P2	*ips*	2.10	0.21	2.11	0.20	2.05	0.37
	cont	2.10	0.21	2.08	0.24	2.16	0.30
ACoA		1.47	0.17	1.63	0.43	−	−
PCoA	*ips*	1.45	0.31	2.01	0.39	1.74	0.39
	cont	1.45	0.31	2.08	0.24	1.66	0.58
BA - B1		3.17	0.51	2.77	0.40	2.93	0.36
BA - B2		3.29	0.44	2.86	0.56	3.31	0.34
ICA		4.72	0.26	4.65	0.44	4.70	0.51

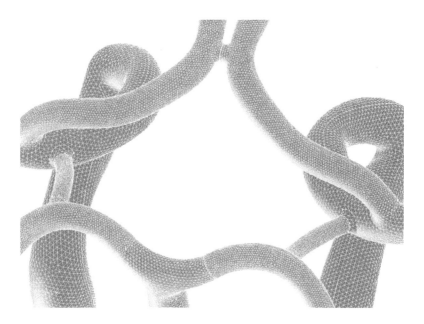

Figure 5.8: An example of the tetrahedral element mesh created for the idealized circle of Willis models.

5.3 Patient Specific Geometry

The idealized CAD models are useful for examining the effect of variations in arterial diameter on blood flow throughout the circle as the diameters can be parametrically driven. However, this technique only works well for small variations in diameter and requires a separate CAD model for every type of anatomical variation. It was decided that to further the research, rather than focus on the creation of idealized models, it would be necessary to develop a technique which allowed for rapid generation of patient specific models, of sufficient quality to use in a CFD simulation.

The TOF scans used to generate the raw data used for processing were performed on a *General Electric (GE)* 1.5T scanner, where the important scanning parameters are listed in Table 5.2. The result of the scan is a series of 2D images, each of which corresponds to an transverse location in the brain. These 2D images can be read in sequentially to create the 3D array $M(x, y, z)$, where x, y, z correspond to axes normal to the sagittal, coronal and transverse planes respectively. While the standard *DICOM* data format of the 2D images uses 16 bit integers to represent the pixel intensities, these values were linearly interpolated such that their min and max integer values map to the range 0 - 1 as double precision numbers, so that certain mathematical operations could be performed on the data.

Table 5.2: Time of Flight parameters used in the acquisition of the MRI scans. Note: the definition of these parameters is given in Appendix A.

T_R	30ms	Matrix Size	512×512
T_E	3.4ms	Pixel Spacing	0.4mm
NEX	2	Slice Spacing	0.5mm
Flip Angle	$20°$	Number of 3D slabs	4

The approach utilized in the present study for the segmentation of the arterial lumen from the TOF dataset uses a variation of the *marching cubes* algorithm developed by Lorensen and Cline [92]. While the detailed explanation of the algorithm is presented in Appendix B, the basic idea is that given a 3D array $M(x, y, z)$ where each entry in the array is the MR signal intensity from a given *voxel* (Figure 5.9(a)), eight values from two sequential slices (Figure 5.9(b)) can be assigned as the vertices of a hypothetical cube (Figure 5.9(c)) which is 'marched' through M. By comparing the numerical values at the vertices of the cube to a user specified *isosurface value* i_{so}, cases arise where some of the vertices are above i_{so} and some are below and it

can hence be determined that a surface corresponding to the arterial wall will pass
through the cube (Figure 5.9(d)).

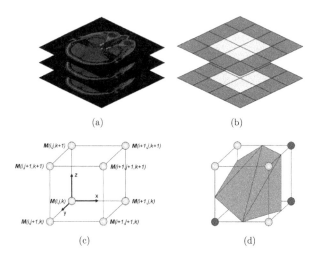

(a)   (b)

(c)   (d)

Figure 5.9: The basis for marching cubes illustrated with (a) a stack of MRI images from
a 3D TOF scan forming the 3D array M(x,y,z), (b) two consecutive slices in the array with
the 4 vertices used per slice highlighted, (c) the 8 array values used to form the vertices
of the cube and (d) and example triangulation through a cube, where light blue and dark
blue represent MR intensity values which are lower and higher than the isosurface value,
or vice versa.

The arterial wall is described by a surface triangulation, which is the collection
of all of the resulting triangles as the cube is marched through the array and is
defined by a *Face* array F and a *Vertex* array V:

$$
F = \begin{vmatrix} V_1 & V_2 & V_3 \\ V_1 & V_3 & V_4 \\ V_2 & V_3 & V_5 \\ & . & \\ & . & \\ & . & \end{vmatrix} \qquad V = \begin{vmatrix} x_1 & y_1 & z_1 \\ x_2 & y_2 & z_2 \\ x_3 & y_3 & z_3 \\ x_4 & y_4 & z_4 \\ x_5 & y_5 & z_5 \\ & . & \\ & . & \\ & . & \end{vmatrix} \qquad (5.1)
$$

where V stores a list of spatial coordinates with each row representing a vertex of one or more triangles and each row in F stores the pointers to the rows of V which comprise the 3 vertices of a given triangle. While the spatial coordinates of the vertices generated by the marching cube algorithm will lie somewhere within the dimension of the MR dataset (L, M, N), the x, y, z components may be multiplied by the pixel x and y and slice spacing respectively (Table 5.2) so that a scaled surface triangulation results.

There are a number of complications that must first be addressed before the marching cubes algorithm can generate a suitable surface triangulation. The first major problem is the presence of extra blood vessels in the dataset or MR signal from other tissues (Figure 5.10(a)), resulting in the generation of unwanted surface topology (Figure 5.10(b)). The second major problem is that frequently a very low MR signal is generated in the posterior communicating arteries, resulting in low visibility in the scan (Figure 5.10(c)). While most of the noise in the dataset, such as signal from tissues other than the lumen can be removed with various filtering algorithms, the problem arises that the posterior communicating arteries are of a similar size and intensity as much of the noise itself. As a result, filtering algorithms may remove the communicating arteries in the filtering process. Furthermore, designing an algorithm that will filter out noise around the communicating arteries suffers from robustness problems because of the issue that the shape and size and even the presence of these arteries varies considerably between individuals.

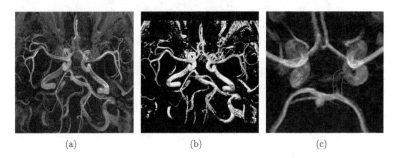

(a)   (b)   (c)

Figure 5.10: Problems associated with directly segmenting the circle of Willis surface topology from the MRI data. (a) the presence of unwanted blood vessels and other tissues in the dataset causing (b) unwanted surface topology in the segmented model (c) the frequent poor visibility of the posterior communicating arteries.

A third problem with the MR data is that due to the limited resolution and presence of noise in the scan, the arterial wall generated from the segmentation frequently appears coarse with 'lumps' in the surface (Figure 5.11(a)). While various smoothing algorithms can be used on the dataset to remove the lumps (Figure 5.11(b)) and interpolation algorithms used on the dataset to effectively divide the voxels up and produce a finer surface topology (Figure 5.11(c)), the variation in size of the arteries comprising the circle of Willis makes it difficult to design algorithms to enhance the quality of the surface topology of one part of the model without reducing it in another part. This is because the smoothing is related to the size of the arteries; adequately smoothing the larger arteries such as the internal carotids tends to remove the smaller arteries, such as the communicating arteries, from the dataset. Conversely, smoothing the smaller arteries adequately has little effect on the larger arteries. Furthermore, care must be taken when smoothing the surface topology so that the effective diameter of the vessels is not altered to such an extent that the amount of blood flow through the vessels would be over or underestimated by the fluid simulations.

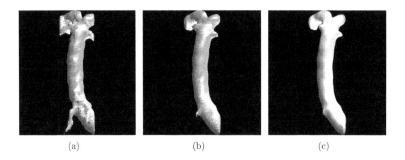

            (a)   (b)   (c)

Figure 5.11: (a) Lumps in the arterial wall surface and coarseness resulting from the limited resolution and noise in the MRI scan, (b) the result of smoothing the dataset to remove the lumps and (c) the result of interpolating the dataset to achieve a finer surface.

The common feature in the problems outlined thus far is the difficulty in automating the various filtering algorithms to work effectively on the different parts of the dataset. However, it is relatively simple for anyone with a knowledge of the cerebral vasculature to differentiate between noise and a desired part of the model, to locate a posterior communicating arteries and to determine which are the larger arteries and which are the smaller arteries. To this end it was decided that the best way to develop the technique was to let the algorithms be *interactive*, so that the user specifies regions of interest in which they can either remove noise, smooth the arterial wall or enhance the smaller communicating arteries etc.

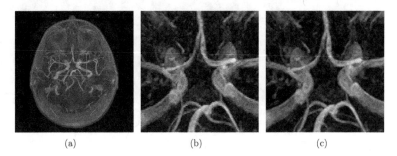

(a)   (b)   (c)

Figure 5.12: A transverse MIP of the circle of Willis illustrating (a) a rectangular ROI drawn around the circle of Willis, (b) the dataset reduced to the rectangular ROI and (c) the dataset subsequently interpolated, effectively dividing up the voxels present in the original dataset.

The remainder of this discussion will outline the various features of the interactive software, illustrating by example how they were used to create 3D models suitable for a CFD simulation. The software was developed using *MATLAB* for the reason that it provided a large range of useful functions that could be easily modified to suit the present needs, such as readers for the DICOM data format as well as a range of visualization functions pertaining to the creation of GUI based tools. This range of features, combined with the interpretive nature of the *MATLAB* programming language provided an ideal environment for quickly and easily experimenting with algorithms and techniques for enhancing the MR data.

The key feature of the software is the ability to draw a *region of interest (ROI)* in the shape of a rectangle (Figure 5.12(a)), an arbitrary polygon or a freehand sketch, and perform operations within these regions. The ROI's are used to create a *binary image* of the same size as the dataset (from whichever orientation the dataset is being viewed from) with a logical 0 outside the ROI and a logical 1 inside. The mapping of this binary image to the 3D array can be used to ensure that only the array entries which correspond to the ROI are effected by the data enhancing operations. The 3D MR dataset can be viewed from the three orthogonal transverse, sagittal and coronal views outlined in Chapter 1, as either a *maximum intensity projection (MIP)* or as individual slices. The standard procedure for creating a segmented model would typically involve initially drawing a rectangular ROI around the circle of Willis (Figure 5.12(a)) and *reducing* the dimensions of the dataset (L, M, N) to

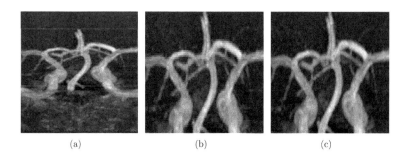

(a)   (b)   (c)

Figure 5.13: A coronal MIP of the circle of Willis illustrating (a) a rectangular ROI drawn around the circle of Willis, (b) the dataset reduced to the rectangular ROI and (c) the dataset subsequently interpolated, effectively dividing up the voxels present in the original dataset.

a smaller array (L', M', N') (where $L' < L, M' < M, N' < N$) so that only this part of the MR dataset would be stored in memory and enhanced in further operations (Figure 5.12(b)). The dataset as a whole could then have *interpolating* algorithms applied to it, increasing the number of voxels in the dataset, in order to produce a finer surface topology (Figure 5.12(c)). It should be noted that this is the only operation developed not requiring a specified ROI. Furthermore, the interpolating operation does not have the effect of removing lumps in the surface triangulation of the arterial wall and may in fact accentuate them in some cases. Increasing the size of the dataset around the circle of Willis only allows for a finer surface topology in the sense that it will be composed of a greater number of triangles. As illustrated in Figures 5.13(a) - 5.13(c), the procedure of reducing and interpolating can be applied in any of the three orientations.

With only the required part of the dataset present, the next steps would generally involve removing unwanted noise from the dataset and enhancing the arterial lumen. The lumen of the entire circle of Willis encompasses a range of intensity values, which overlap with the noise in the image such that the actual amount of noise present in the final isosurfaced model depends upon the isosurface value required to segment the entire circle of Willis. To streamline the process of determining the correct isosurface value in order to segment the circle of Willis, the isovalue could be dynamically altered and the data within the 3D array below the current isosurface value removed from the display (Figure 5.14(a)). This displayed image would correspond to placement of the arterial wall and it can be observed in Figure 5.14(a)

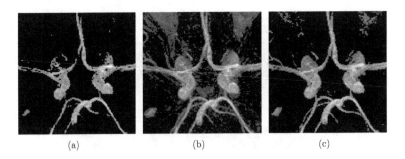

Figure 5.14: A transverse MIP illustrating the effect of a dynamically altering isosurface value (a) a value too low to segment all of the required geometry (b) an isosurface value which is too low and causes a significant amount of noise to be segmented but also places the arterial wall in the wrong location and (c) an isosurface value which is appropriate to segment all of the circle of Willis, placing the arterial wall in the correct location and segmenting an acceptable level of noise.

where the isovalue is too high, that only a small portion of the circle of Willis would be segmented. Conversely, if the isovalue was set too low then a greater amount of noise will be segmented, but more importantly the resulting surface triangulation may be placed away from the lumen of an artery, meaning that the arteries appear larger in 3D model than is the case anatomically (Figure 5.14(b)). Generally the isosurface value that illustrates the correct placement of the arterial wall would also include the segmentation of a significant amount of noise (Figure 5.14(c)). The important point is that the regions that would need to be targeted for noise removal could be observed and the next step in the data processing would generally involve removing unwanted noise from the dataset by selecting ROI's around the circle of Willis (Figure 5.15(a)) and *deleting* the data (Figure 5.15(b)). Mathematically this is equivalent to setting the voxel intensity to zero as:

$$M\left(i, j, k\right)_{ROI} = 0 \tag{5.2}$$

where i, j, k correspond to array entries located within the ROI. It should be noted that the data deleted from within the ROI in Figure 5.15(b) was removed through *every* transverse slice (or every coronal or sagittal plane if deleting data from the coronal or sagittal views respectively). In order to produce very fine operations, this deleting technique was extended to form an *eraser* tool which allowed the user to delete single voxels as the cursor was moved over them.

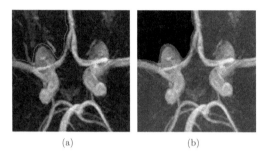

(a)   (b)

Figure 5.15: A transverse MIP illustrating (a) a freehand ROI drawn around some unwanted data and (b) the data deleted from the MR dataset.

In situations where it is required to delete data from only certain slices, two
options were created. The first option was to *split* the dataset by selecting a ROI
around the part of the dataset required to be split (Figure 5.16(a)) and subsequently
performing operations only on that visible split portion (Figure 5.16(b) - 5.16(c)).
The second option was to view individual slices and by selecting ROI's within the
slice (Figure 5.17(a)) only remove data within that slice (Figure 5.17(b)).

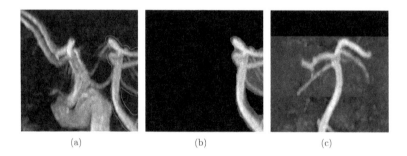

(a)   (b)   (c)

Figure 5.16: A sagittal MIP illustrating (a) a freehand ROI drawn around the posterior
of the CoW, (b) the data within the ROI separated from the rest of the MR dataset and
(c) a coronal view of the separated region of the MR dataset.

The other major advantage of splitting the dataset is the ability to separate out
arteries of different sizes so that each artery can be smoothed to different degrees.
As an example of this, the two anterior cerebral arteries have been selected from a

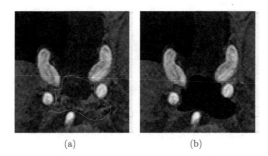

(a)   (b)

Figure 5.17: A transverse slice illustrating (a) a freehand ROI drawn around some unwanted data and (b) the data deleted from the MR dataset.

transverse ROI (Figure 5.19(a)) and the dataset split and switched to a coronal view
(Figure 5.19(b)). By then selecting another ROI and smoothing the data within the
ROI, only these arteries will be affected by the smoothing operation (Figure 5.19(c)).
Mathematically, the operation of smoothing is accomplished by the convolution of
the 3D array with a *gaussian kernel G*:

$$M\left(x,y,z\right)_{smooth} = M\left(x,y,z\right) * G\left(x,y,z,\sigma\right) \tag{5.3}$$

where the gaussian kernel is defined as:

$$G\left(x,y,z,\sigma\right) = \frac{1}{\sqrt{2\pi\sigma^2}}e^{-\frac{x^2+y^2+z^2}{2\sigma^2}} \tag{5.4}$$

and is illustrated as a function of only one spatial dimension (for simplicity) in
Figure 5.18. The width of the gaussian function is determined by the parameter σ,
which can likened to the *smooth level* and must be 'tuned' to the size of the artery
being smoothed. The gaussian kernel is a 3D cubical array of size $6\sigma + 1$ (since
the function is essentially zero greater than a distance 3σ from the origin) and is
evaluated at the discrete integer values corresponding to i, j, k position within the
array, treating the center of the array as the origin. The convolution is performed

computationally as:

$$M\left(i,j,k\right)_{smooth} = \sum_{l=1}^{6\sigma+1}\sum_{m=1}^{6\sigma+1}\sum_{n=1}^{6\sigma+1} M\left(i+l-1,j+m-1,k+n-1\right)G(l,m,n) \quad (5.5)$$

Which yields a smoothed matrix of size $(L+6\sigma, M+6\sigma, N+6\sigma)$ that must be 'clipped' back to its original size following the convolution.

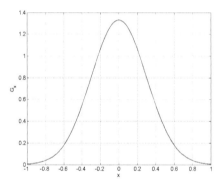

Figure 5.18: The Gaussian function in 1D.

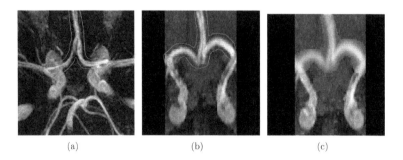

(a)   (b)   (c)

Figure 5.19: A transverse MIP illustrating a freehand ROI around the ACA's, (b) a coronal MIP of the separated data with another freehand ROI drawn around the ACA's and (c) the smoothed dataset within the two ROI's. Note: the smooth level has been exaggerated for effect.

As mentioned previously, the poor visibility of the posterior communicating arteries presents a problem in the model generation. Given that the marching cube type algorithm requires a single isosurface intensity value in order to extract the surface topology, the intensity of the posterior communicating arteries must be equal to or greater than this intensity to become part of the model. Simply reducing the isosurface value is of no advantage since other unwanted noise in the data will then start to be segmented too. What is required is a way to selectively enhance these arteries. The typical procedure for enhancing these posterior communicating arteries would involve selecting a ROI around the artery (Figure 5.20(a)) and splitting the dataset (Figures 5.20(a) - 5.20(c)). By subsequently selecting another ROI around the artery from a perpendicular view (Figure 5.20(e)) the ROI may have a brightening algorithm applied to it enhancing its visibility (Figure 5.20(f)). Mathematically the brightening algorithm takes the form:

$$M\left(i,j,k\right)_{ROI} = M\left(i,j,k\right)_{ROI}^{\gamma} \qquad (5.6)$$

where γ is the *brightening level*. This operation means that no intensity value can be greater than 1 and if γ is chosen as less than 1, the effect is to decrease contrast by brightening the darker regions in the ROI, or conversely if γ is greater than 1, the effect is to increase contrast by darkening the darker regions. Figures 5.20(d) - 5.20(f) illustrate another feature developed for enhancing the MR data, namely the ability to switch the colourmapping from greyscale to an *RGB* intensity scale. The reason for switching colourmaps is that variations in colour around the edge of an artery are much more obvious than changes in greyscale intensity. In many operations which involve placing a ROI in close proximity to wherever the arterial wall is determined to be, it can be beneficial to visualize the dataset in this manner.

To reiterate, the dataset enhancing operations include reducing, interpolating, deleting, brightening, and smoothing the data and this may be done with either a greyscale or RGB colourmap, viewing the data from either a transverse, sagittal or coronal orientation, as either a MIP or as individual slices. While the usual procedure outlined generally involved reducing the dataset first, then interpolating, then deleting, then brightening, then finally smoothing, these operations can be performed in any order. Furthermore, the dataset can be segmented using the isosurface algorithm between operations in order to visualize the effect on the final model.

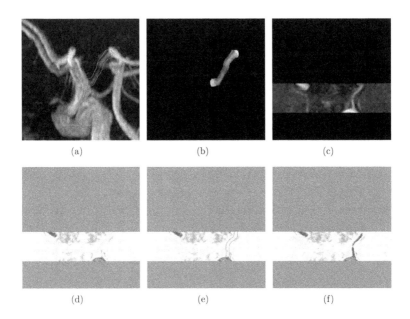

Figure 5.20: A sagittal MIP illustrating (a) the RPCoA selected with a freehand ROI, (b) the RPCoA separated from the rest of the dataset, (c) a transverse MIP of the separated RPCoA, (d) the colourmap changed from greyscale to RGB, (d) a second freehand ROI drawn around the RPCoA and (d) the result after applying a brightening algorithm.

It should be noted at this point that there is a significant body of work in the field of blood vessel enhancement, often using the approach of convolving the 3D array with a *Laplacian of Gaussian kernel* in order to determine a *Hessian matrix* for every voxel within the array [48, 131]. Based on the eigenvalues of the Hessian matrix the likelihood of a given voxel being a part of a tubular structure (a blood vessel) can be determined. A filter of this type was included in the software based on the work of [48], but it was found that the filter did not significantly enhance the communicating arteries (which is essentially the biggest problem in isosurface generation) and the amount of computational time required meant that a trained user could produce the desired surface topology in a shorter period of time via a combination of the aforementioned algorithms.

Although the deleting operation just outlined allows for the removal of unwanted noise from the dataset, it is still possible that unwanted pieces of blood vessel or noise become segmented, producing closed surface triangulations at various locations around the circle of Willis triangulation (Figure 5.10(b)). This may be the result of an inability to isolate the unwanted data or a failure to notice it when inspecting the surface triangulation. In either case, to remove this noise from the surface triangulation so that there is only *one* closed surface forming the circle of Willis, a specialized *neighbour painting* filter was developed. The idea behind the filter is that given a dataset where there are other surfaces around the circle of Willis (Figure 5.21(a)), the user can interactively select a triangular facet on part of the surface triangulation which forms the circle of Willis (Figure 5.21(b)). Using the connectivity information of the triangulation, faces which are connected to this initial face can be located and then subsequently faces connected to these faces in a recursive manner such that the entire surface topology of the circle is traversed, but any unconnected noise will not be.

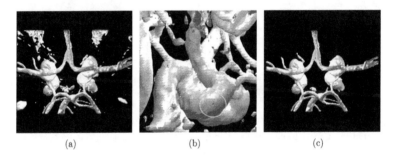

(a)   (b)   (c)

Figure 5.21: A surface triangulation of the arterial wall of the circle of Willis illustrating (a) unwanted noise in the triangulation, (b) a triangular facet selected on the carotid siphon (highlighted within blue circle) and (c) the noise removed with the neighbour painting filter.

As illustrated in Figure 5.22, the filter works by initially selecting one of the vertices of the selected face which becomes the starting entry in a list of all the vertices connected to the circle of Willis. The algorithm then begins searching the *Face* array for any other faces which use that vertex and if one more faces are found then they are *flagged*, by storing the row position of that face in another array. The algorithm then loops over all of the faces that were found in the search and examines the vertices of each face to see if they have been covered previously (have been flagged in the same manner as entries in the *Face* array are) and if not then these vertices are added to the list of connected vertices. The algorithm increments the position in the list of connected vertices and repeats until it comes to the end of the list, meaning that *all* and *only* the faces comprising the circle of Willis have been covered. The list of stored faces can then be used to remove any unwanted entries from the Face array F.

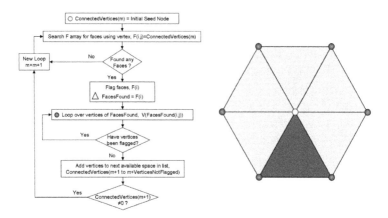

Figure 5.22: The implementation of the neighbour painting filter algorithm with the geometrical entities highlighted by an example.

As mentioned previously, the generation of a surface topology is only a part of creating a model suitable for a numerical fluid simulation, as the boundary faces and the porous blocks must be defined. As long as all of the afferent and efferent arteries extend to the ends of the 3D dataset, when the isosurfacing algorithm is applied to the dataset the termination will be left hollow (Figure 5.23(a)). It is then possible to create a *boundary* array B for each of the six faces of the cube formed by the 3D array where one level contains the values of the matrix M at the boundary and another level is padded with zeros. As an example to illustrate, the boundary matrix for the right sagittal plane (where the RMCA is truncated) would take the form:

$$B\left(i, 1, k\right) = M\left(i, j_{end}, k\right) \tag{5.7}$$

$$B\left(i, 2, k\right) = 0 \tag{5.8}$$

where j_{end} represents matrix entries at the maximum sagittal value. By then applying the marching cubes algorithm to this 3D dataset, the result is a separate surface triangulation of the end plane of the arteries. As long as the same isosurface value is used in both M and B, this surface triangulation will directly map onto the end of the artery (Figure 5.23(b)). The porous blocks can be created in much the same manner except that the porous block dataset P is formed by essentially extruding the values of the matrix M at the boundary with no zero padding. Continuing the example of the right sagittal plane, P would take the form:

$$P(i, j_{all}, k) = M(i, j_{end}, k) \qquad (5.9)$$

where j_{all} represents the number of matrix entries (geometrically equivalent to the length of the extruded porous block) which for the present study was chosen to be one eighth of the dataset in the particular dimension (e.g. if M had 200 entries in the j direction then j_{all} in P would be 25). By again applying the marching cubes algorithm with the same isosurface value, the porous block will map directly onto the end of the efferent artery (Figure 5.23(b)). Finally the faces at the distal end of the porous blocks (Figure 5.23(c)) are created by creating a copy of the initial boundary face and translating its coordinates the required direction by adding (or subtracting in some cases) the length of the porous block so that it directly maps onto the distal end of its porous block.

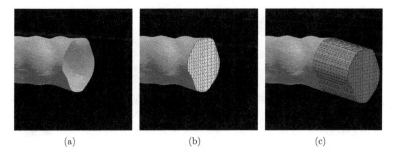

(a)   (b)   (c)

Figure 5.23: The right MCA of a circle of Willis model illustrating (a) the hollow artery, resulting from the isosurfacing of the main dataset, (b) the interface surface triangulation created by the isosurfacing of a boundary matrix and (c) the porous block and efferent outlets created by the isosurfacing of a porous block dataset and copying and translating the triangulation interface respectively.

The final piece of geometry to be created are the arterial stenoses. Physiolog-

ically, the primary site for stenosis formation is at the carotid bifurcation, where
the common carotid artery splits into the internal and external carotid arteries, but
also in the vertebral arteries to a degree. Since these regions have been truncated
from the model (due to limitations of computer resources and MRI scanning time),
a different manner of including their effects is necessitated. It was decided to keep
the same afferent terminations (Figure 5.24(a)), and apply the same approach for
extruding the porous blocks to the afferent arteries, instead extruding *stenosis blocks*
(Figure 5.24(b)), to which surface triangulation could be deformed, to mimic the
effects of a stenosis. The length L of the stenosis blocks was chosen to be half the
dataset size in the transverse direction, which was approximately 150mm for most
models.

Since all three stenosis blocks are extruded as a single surface triangulation,
the neighbour painting filter was incorporated as a means to interactively select a
given stenosis block and separate its faces and vertices from the other blocks (Figure
5.24(c)). It should be noted at this point that the afferent arterial terminations are
not perfectly circular and therefore the stenosis blocks are not perfectly cylindrical.
Deforming the surface triangulation hence requires a method which does not depend
upon modifying a circular cross section but will work for any shape. The method
devised for each stenosis block began by determining the centroid of the stenosis
block, which is a problem reduced to determining the arithmetic mean of all of its
vertices $\mathbf{V}_i$:

$$\mathbf{V}_c = \frac{1}{N_{verts}} \sum_{i=1}^{N_{verts}} \mathbf{V}_i \qquad (5.10)$$

where N_{verts} is the number of vertices in the triangulation of the single stenosis
block and $\mathbf{V}_c$ are the x_c, y_c, z_c coordinates of the centroid, where the calculation
was performed on each of the three x, y, z components separately. Following this
calculation, the x and y coordinates of the triangulation were shifted to conform to
an exponential function (similar to the gaussian function) and chosen to be:

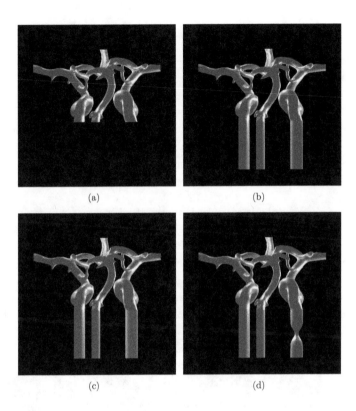

(a)   (b)

(c)   (d)

Figure 5.24: (a) A coronal view of the circle of Willis triangulation, (b) the stenosis blocks extruded from the afferent terminations of the model, (c) a single stenosis block separated and highlighted and (d) the vertices of its surface triangulation deformed to simulate a stenosis.

$$x_{i,new} = x_i + DoS \left(x_c - x_i \right) e^{-\alpha \left(z_i - z_{\frac{L}{2}} \right)^2} \qquad (5.11)$$

$$y_{i,new} = y_i + DoS \left(y_c - y_i \right) e^{-\alpha \left(z_i - z_{\frac{L}{2}} \right)^2} \qquad (5.12)$$

where DoS is the *degree of stenosis* (describing the percentage reduction in area of the vessel at its most narrow point), $z_{\frac{L}{2}}$ is the transverse coordinate midway along the length of the stenosis block, and α is a parameter controlling the taper of the vessel (similar to σ in the gaussian function) which was arbitrarily set to be 0.15. The effect of the stenosis is to cause a loss in blood pressure, much greater than the normal gradual loss in blood pressure as blood travels through the cardiovascular system. It is important to note that in placing the stenosis much closer to the circle of Willis, the assumption is made that the loss in afferent blood pressure at the start of the computational model would be the same as if the carotid bifurcation were included in the model and the stenosis was placed there. Furthermore it is assumed that the stenosis is still far enough away from the communicating arteries and the bifurcations of the internal carotid and basilar arteries that any secondary flow distal to the stenosis will not have any effect on the flow patterns through them.

To recap slightly, an example of the raw MRI data for a complete circle of Willis model is illustrated in Figure 5.25(a) with the final segmented surface topology illustrated in Figure 5.25(b). All unwanted noise has been removed, the arteries enhanced and the boundary faces and porous blocks generated. The model was then exported to the grid generator in a *stereo lithography (STL)* format with a separate file generated for the circle of Willis arterial wall, the afferent inlets, porous block interfaces, the porous block and stenosis block walls, and the efferent outlets. The overall time period for creating such a model depends upon the quality of the scan, but would generally take approximately 30 minutes on a desktop PC.

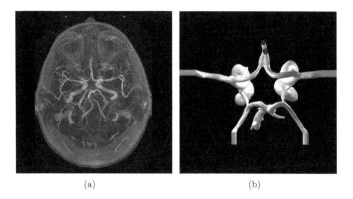

(a)   (b)

Figure 5.25: (a) A transverse MIP of the raw MRI data (b) the final segmented model.

For the present study a total of five MRI scans were performed on healthy volunteers. Of the five scans, three anatomical configurations were chosen for investigation, namely a complete circle of Willis (Figure 5.25(b)), a unilateral missing

PCoA (Figure 5.26(a)) and fused ACA configuration (Figure 5.26(b)). Referring
to the anatomical studies outlined in Chapter 1, the fusion of the ACA's over a
short distance is among the more rare variations occurring within the population.
The absence of a PCoA is also an extremely rare variation, but given the use of
MRI to image the cerebral vasculature, it is more likely that this individual in fact
possesses only a hypoplastic PCoA. In terms of the capacity to provide collateral
circulation, a hypoplastic PCoA will not provide any significant contribution and it
is considered that in terms of the CFD simulation the artery can remain omitted
without reducing the validity of the results significantly.

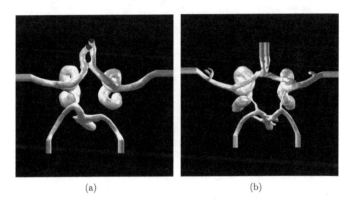

(a)   (b)

Figure 5.26: Axial views of the (a) missing PCoA and (b) fused ACA patient specific
circle of Willis anatomical variations created.

Generating the computational grids for the segmented geometries poses a greater
challenge compared to the CAD models, the reason being that only a surface triangulation is specified. While CAD data contains essentially all of the edges and faces
used in the creation of the model. As a result it is relatively easy to define faces
with which to split the geometry back into simpler geometric entities such as single
arterial segments and junctions, thereby easing the burden on the mesh generator.
With the segmented data this is not the case and to circumvent the problem the
mesh generator *Harpoon* was utilized to create an *adaptive cartesian* mesh. The
adaptive cartesian meshing technique generates quality meshes on geometries of essentially any complexity and works well with a surface triangulation description of
the model, making it ideal for the segmented circle of Willis geometries. The technique generates mainly hexahedral elements in the interior of the model, but also
tetrahedral, wedge and pyramid elements at the surface. The meshes generated on

the segmented models consisted of a total of approximately 2 million elements which were found to give a mesh independent solution (Chapter 9).

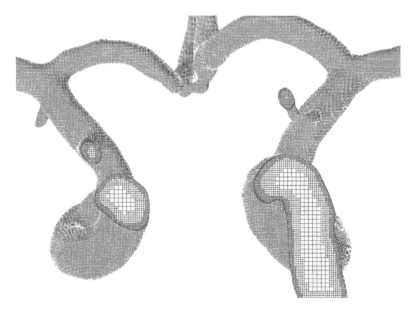

Figure 5.27: An example of the adaptive cartesian mesh created for the segmented circle of Willis models.

With unstructured meshing techniques (to which both the TGrid scheme and adaptive cartesian meshing belong), generally there are four basic types of 3D cell which may be produced; tetrahedral, hexahedral, wedge and pyramidal cells (Figure 5.28) , all of which may vary in size and shape and be used in various combinations within a domain. With the TGrid scheme applied to the idealized circle of Willis geometries, only the first of these four types was utilized in the mesh, namely the tetrahedral cell. With the adaptive cartesian meshing technique however, the majority of the cells are hexahedra, located in the centre of the domain, but all of the other three types of cell are generated around the wall of the model.

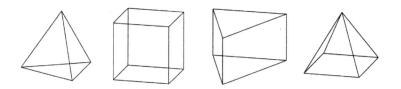

Figure 5.28: Tetrahedral, hexahedral, wedge and pyramid cell types used in unstructured grids.

Chapter 6

Cerebral Autoregulation Modelling

6.1 Introduction

As was outlined in Chapter 2, the body has a number of physiological control systems, designed to regulate blood flow by the alteration of either blood pressure or peripheral resistance. Clinically however, blood pressure is the physicians *control variable* (i.e. a patients blood pressure can be set as desired by the administration of certain drugs), so there is little to gain in attempting to model the physiological mechanisms which regulate blood pressure (i.e. the baroreceptor and chemoreceptor loops and the functioning of the kidney) as they can be overridden. Peripheral resistance on the other hand, generally remains under the body's control and it is therefore important to be able to understand and model its function. When considering the cerebral vasculature the most important mechanism is cerebral autoregulation, and to this end, part of the present work has been dedicated to developing mathematical models of autoregulation with the intention of describing (at least in part) the physiology of the autoregulation process and using parameters which may be obtained from experimental data on humans. To aid in the future goal of utilizing the modelling technique as a clinical tool, a desirable feature was to compose the model of mathematical variables which have a direct physiological significance rather than use abstract variables which may produce the correct result but lack any physiological meaning.

Autoregulation is accomplished by the alterations in smooth muscle tone in the tunica media of the smaller arteries on the surface of the brain. With the efferent termination of the 3D circle of Willis geometries immediately downstream of the origin of the major cerebral arteries (for the reasons mentioned previously), the inclusion of smooth muscle tone and/or arterial diameters can only be incorporated as mathematical variables if some sort of simplified model of the arterial network distal to the circle of Willis is adopted. There have been a number of studies adopting such an approach (for example [63, 115, 154]) where 1D electrical network analogies can be combined with a Poiseuille flow relation [169] between arterial resistance and diameter and Laplace's law [1] relating smooth muscle tone to arterial diameter. This approach has numerous advantages, including modelling autoregulation in the region of the cerebrovascular network where it is actually occurring, overcoming the problem of the large number and small size of these arteries, and yielding the changes in smooth muscle tone and hence arterial diameter throughout the cerebral vasculature in response to changes in different physiological parameters such as mean arterial blood pressure. The major drawback of this approach is that there exists limited data regarding the diameters, lengths, wall compositions, numbers and variability of smaller arteries of the cerebral vasculature in humans, especially under pathologic conditions. For this reason results from experiments on animals tend to be used in order to specify the large number of parameters required with this modelling approach. In a clinical environment this level of detail is neither required nor desired and useful results can be generated by use of lumped parameter models in order to represent the distal arterial networks, many of which have been implemented in other studies [54, 74, 146, 158]. For this reason it was decided to keep cerebrovascular resistance CVR as the key mathematical variable which, as will de outlined in Chapter 7, can couple in elegantly with the incorporation of the porous blocks at the efferent model outlets.

Modelling the driving force for smooth muscle relaxation or contraction (hence an alteration in CVR) poses just as great a challenge as modelling its distribution throughout the cerebrovascular network. While general components of the overall process, such as the metabolic, myogenic and nervous components discussed in Chapter 2 and a number of their chemical pathways are known, the complete picture of all of the physiological processes and chemical pathways involved and their interaction has not yet been fully elucidated. Furthermore, the knowledge of certain chemical pathways alone is not sufficient basis for a mathematical model, as there are many key parameters relating to the relative contribution of each component which are also at present unknown. While the relative importance of each component could potentially be circumvented if the complete interaction between all of the chemical pathways was understood, at present this remains a limitation. Many studies have sought to include all aspects of autoregulation by categorizing each component and developing separate models [154], but suffer from the drawback of not necessarily capturing the chemistry behind each component and more importantly, not having sufficient information to determine the relative contribution of each component. Recent work [17] has made an attempt at describing a sizeable portion of the blood chemistry, brain metabolism and smooth muscle mechanics, incorporating it within a 1D network model of the cerebral vasculature, but suffers from the drawback of having an incredibly complex system of equations, requiring vast numbers of parameters.

For the present study it was decided to implement a simplified description of the overall autoregulatory process, requiring fewer parameters than more complex models, but which may be set using experimental data from humans. For the initial work with the idealized circle of Willis geometries an autoregulation model was adapted from other work, and subsequently a new model was developed for incorporation with the patient specific circle of Willis geometries. The new model included a more complex model of the blood chemistry and brain metabolism such that it may provide a slightly more accurate picture of the catalysts for autoregulation, as well as increase the scope of how the model may be used clinically and for validation purposes.

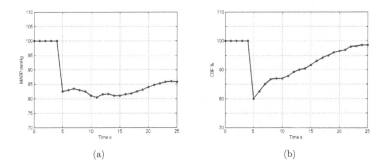

Figure 6.1: The experimental data of Newell et al [112] showing (a) the reduction in $MABP$ brought about by a thigh cuff manoeuver and (b) the its effect on CBF, illustrating the return to normal by a reduction in CVR.

One of the important features of the autoregulation mechanism is the transient dynamics, indicating the timescale for the physiological processes to occur in order for CVR to adjust to maintain blood flow. In the study of Newell et al [112] a group

of volunteers were subjected to a rapid step change in arterial blood pressure by
the prolonged inflation, then rapid deflation of blood pressure cuffs placed around
the thighs. During the experiment arterial blood pressures were monitored with
intra-arterial catheters and transducers, and blood flow in an ICA monitored by
an electromagnetic flowmeter. The results averaged among the volunteers for the
arterial blood pressure (Figure 6.1(a)) and ICA flowrate (Figure 6.1(b)) as a function
of time clarify two important features of the autoregulation mechanism; the first
being that the time scale is of the order of about 20 seconds for the CVR to decrease
to a level that *cerebral blood flow CBF* is restored, and secondly that CVR exhibits
first order dynamics. This result has been incorporated into both autoregulation
models.

6.2   Idealized Circle of Willis Autoregulation Model

The cerebrovascular resistance dynamics implemented in the autoregulation model
for the idealized geometries were based on a model used by Ferrandez et al [37, 46],
in which CVR is described by a first order ode:

$$\frac{dCVR}{dt} = G_P \left(CBF - CBF_{SP} \right) + G_I \int_0^T \left(CBF - CBF_{SP} \right) dt \qquad (6.1)$$

where CBF_{SP} is a *set point* or 'normal' reference flowrate. The first term on the
right hand side of (6.1) represents a *proportional* controller where the time rate
of change of CVR is proportional to deviation in flowrate from its set point value
$(CBF - CBF_{SP})$ at any given time and has a control gain G_P (taken as 70). The
second term on the right hand side of (6.1) represents an *integral* controller, with a
control gain G_I (taken as 0.1), where the rate of change of CVR also depends upon
the integral of the flowrate error over time. The proportional control term has a
reasonably clear physiological meaning in that when the blood flow through a given
efferent artery deviates from what it would be under normal homeostatic conditions,
CVR will hence alter accordingly. The integral control term has less physiological
meaning but is incorporated because it acts as a *low pass filter* meaning that CVR
will not react to higher frequency fluctuations in CBF (such as the cardiac cycle)
but rather to changes in the mean value of CBF. Furthermore the integral control
ensures that the flowrate error is always driven to zero.

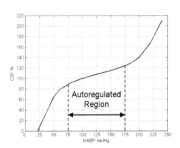

Figure 6.2: Autoregulation curve used in the cerebral autoregulation model of Huntsman et al [77].

In order to incorporate some physiology into the autoregulation model regarding blood chemistry and brain metabolism, a model of Huntsman et al [77] was adopted, based chiefly around the autoregulation curve of Figure 6.2. The model begins with two well known clinical relations, the first of which states that:

$$MABP - MVBP = CVR \cdot CBF \qquad (6.2)$$

where $MVBP$ is the mean venous blood pressure. The second equation relates to brain metabolism stating that:

$$CMRO_2 = CBF \left(C_aO_2 - C_vO_2 \right) \qquad (6.3)$$

where $CMRO_2$ is the cerebral metabolic rate of oxygen consumption and C_aO_2 and C_vO_2 the concentrations of oxygen in arterial and venous blood respectively. The assumption completing the system of equations is that CVR is a linear function of the venous oxygen concentration:

$$CVR = CVR_{SP} \left(1 + A \cdot C_vO_2 \right) \qquad (6.4)$$

where CVR_{SP} is a set point or reference resistance under normal homeostatic conditions and A is a constant scaling parameter. By combining (6.2), (6.3) and (6.4), the resulting relation for CBF can be obtained:

$$CBF = \frac{1}{1 + A \cdot C_a O_2} \left(A \cdot CMRO_2 + \frac{MABP - MVBP}{CVR_{SP}} \right) \qquad (6.5)$$

which makes CBF a function of the metabolic rate and arterial oxygen concentration as well as arterial blood pressure. The reason for generating this relation is that it is in keeping with experimental evidence [55] that cerebral blood flow will increase if the metabolic rate increases, or if the arterial oxygen concentration decreases and vice versa. The determination of the scaling parameters A and CVR_{SP}, initially uses the fact that in the autoregulated region (Figure 6.2) the slope of the autoregulation curve is a linear function of $MABP$ and if both CBF and $MABP$ are normalized by their set point values, the linear function becomes:

$$\frac{CBF}{CBF_{SP}} = \frac{2}{3} + \frac{1}{3} \left(\frac{MABP - MVBP}{MABP_{SP} - MVBP_{SP}} \right) \qquad (6.6)$$

Normalizing (6.5) by dividing it by CBF_{SP} and then equating constant terms and coefficients of $MABP - MVBP$ between (6.5) and (6.6) yields two equations for the scaling parameters, which involving normal or set point values for all of the other variables:

$$A = \frac{2CBF_{SP}}{3CMRO_2 - 2CBF_{SP} \cdot C_a O_2} \qquad (6.7)$$

$$CVR_{SP} = \frac{MABP_{SP} - MVBP_{SP}}{CBF_{SP} \left(1 + A \cdot C_a O_2 \right)} \qquad (6.8)$$

It is implicitly assumed in forming the relation between CBF and $MABP$ (as well as the other physiological variables) that these are steady state values as this is the way that autoregulation curves are generated. For this reason it was chosen to incorporate (6.5) as the set point flowrate in (6.1) as:

$$CBF_{SP} = \frac{1}{1 + A \cdot C_a O_2} \left(A \cdot CMRO_2 + \frac{MABP - MVBP}{CVR_0} \right) \qquad (6.9)$$

meaning that in response to alterations in blood pressure, the transient flowrates through the efferent arteries would alter, but also the reference value to which they are trying to return. Each efferent artery has its own value of blood flow under normal conditions CBF_{SP} as well as the resistance (CVR_{SP}) of the corresponding region of the brain being supplied with blood. Since (6.9) and (6.1) are applied independently to each artery this means that they will each have their own values of A and CVR_{SP} to be evaluated using (6.7) and (6.8) respectively. While the levels of $CMRO_2$ and $C_a O_2$ and $MVBP$ were kept constant for all of the simulations performed using this model the values of $MABP$ were chosen for each efferent artery to be the average pressure on the upstream face of the porous blocks. For incorporation into a numerical simulation, Equation 6.1 was discretized using a first order backwards difference for the time derivative and a rectangular numerical integration method for the integral control term:

$$CVR^{t+1} = CVR^t + \frac{1}{\Delta t} \left(G_P \left(CBF^t - CBF_{SP} \right) + G_I \sum_0^T \left(CBF^t - CBF_{SP} \right) \Delta t \right)$$
$$(6.10)$$

The second important feature of the autoregulation model which must be captured along with the dynamics, in order for simulation results to be clinically useful, are the limits of autoregulation. Given a simulation designed to investigate the effects of stenosis or occlusion of one or more afferent arteries, it is highly likely that the reduction in perfusion pressure through the efferent arteries will reduce blood flow to the point such that the autoregulatory mechanism would not be able to restore it completely by vasodilation of the smaller arteries in the brain. At the time of creation of this model, it was decided to use the limit illustrated in autoregulation curve to which the model was fitted (Figure 6.2), such that the upper and lower limits occur at a $MABP$ values of 75 and 175mmHg respectively. Simulations were subsequently tested using a 1D resistor analogy between pressure, resistance and flow and applying the autoregulation model to a single resistor, essentially treating the brain as a single lumped resistance. The simulations then involved setting $MABP$ to these values at the upper and lower limits of autoregulation and then recording the resulting CVR required to restore CBF at this point. These were found to be upper and lower limits of 145% and 50% respectively of the set point for resistance CVR_{SP}, and these were then applied individually to each efferent artery in the full numerical simulations. Referring to the discussion on the clinical mechanics of blood flow in Chapter 2 an important modelling assumption that has been made for both autoregulation models is that any changes in CVR will not affect intracranial pressure and hence the limitations imposed on CVR will be a result only of the maximum vasodilation and constriction to which the cerebral arteries and arterioles are capable.

Since the cerebrovascular resistance dynamic is activated by a deviation in CBF from its set point value, a measure of this set point is required for all of the efferent arteries. The set point flowrates were calculated from a result postulated by Hillen et al [63] that there is a total inflow of 750mL/min into the afferent arteries of the circle of Willis and the cerebrovascular resistance of territories in the brain follows the ratio of 6:3:4 between the regions supplied by the anterior, middle and posterior cerebral arteries respectively. Treating the resistance of the cerebral territories as an electrical network of parallel resistors (Figure 6.3) the blood flow through the efferent arteries can be determined to be 83.4mL/min through each anterior cerebral artery, 166.7mL/min through each middle cerebral artery and 125.0mL/min through each posterior cerebral artery. It was these flowrates which were set as CBF_{SP} for each artery in (6.1), and used to determine the constant A in (6.7) for each artery, calibrating the autoregulation algorithm.

6.3   Patient Specific Circle of Willis Autoregulation Model

The necessity for improvement of the autoregulation model arose from the fact that firstly, there is no physiological reason why CVR in (6.4) should be a function of the oxygen concentration of venous blood. Furthermore, the experiment with which the autoregulation curve in [55] was generated was never described, reducing confidence in its validity in terms of the interpretation of the curve fit to the experimental data and even whether or not the experiment was performed on humans or other animals.

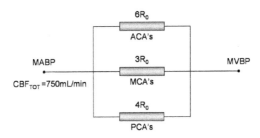

Figure 6.3: A 1D electrical network analogy used to determine the set point flowrates through the efferent arteries of the circle of Willis. Assuming a total inflow of 750mL/min and a resistance ratio of 6:3:4 between the anterior, middle and posterior cerebral arteries respectively.

It was decided that since very little is known about the distribution and relative importance of the myogenic and neurogenic components of autoregulation in the human brain, and in keeping with the goal of incorporating experimental data from humans, their effects would not be incorporated for the present study, leaving their inclusion for future work. The improved model was hence aimed at describing the physiology of the metabolic component of autoregulation.

As was discussed in Chapter 2, metabolism in the brain is mediated by glucose and is therefore performed via aerobic respiration where equal amounts of oxygen and carbon dioxide are consumed and produced in the process respectively. The carbon dioxide produced in the brain tissue diffuses back into the blood stream and is removed by the blood. If blood flow is altered then carbon dioxide levels will either increase or decrease accordingly, depending on the cerebral metabolism. Carbon dioxide undergoes a series of reactions in the blood, one of which is the formation of carbonic acid and bicarbonate. While the decrease in pH associated with the formation of carbonic acid is buffered in the blood, carbon dioxide in the tissue may cause a reduction of pH in the tissue [25], to which there is evidence suggesting a smooth muscle relaxation response [86, 147] possibly mediated by the blocking of ATP specific potassium channels, which thereby leads to a decrease in intracellular calcium levels and hence smooth muscle tone [15, 72, 89]. Nitric oxide is known to be a powerful vasodilator [23, 108], and there is evidence to suggest that it serves a role of modulating alterations in smooth muscle tone [34, 71, 137]. Although the contribution to the overall change in CVR brought about by NO in the larger upstream blood vessels as part of the conducted dilation phenomenon may be much greater than the changes brought about by the metabolite induced dilation, the concept is that it is buildup of metabolites which initiates and facilitates the process. The effects of NO could be thought of as being part of the metabolic component and hence it was decided to not include NO explicitly in the model since firstly, the link between the effects of NO and the buildup of metabolites have not been definitively confirmed, and secondly this link would require a mathematical description of multiple levels of the cerebral vasculature, which was undesirable for the study.

Since the transport of oxygen and carbon dioxide in the blood stream is relatively well understood, it was decided to expand upon this aspect of the previous autoregulation model and include a representation of their transport phenomena, which coupled with the cerebral blood flow would provide the delivery of oxygen and the accumulation of carbon dioxide in the cerebral tissue. The blood chemistry incorporated into the autoregulation model was based on the work of Dash et al [35] and while a complete description of the model will not be given, the basis for the model begins with the biochemical reactions between oxygen, carbon dioxide and hemoglobin in the blood:

$$CO_2 + H_2O \underset{kb'_1}{\overset{kf'_1}{\rightleftarrows}} H_2CO_3 \overset{k''_1}{\rightleftarrows} HCO_3^- + H^+ \quad (6.11)$$

$$CO_2 + HmNH_2 \underset{kb'_2}{\overset{kf'_2}{\rightleftarrows}} HmNHCOOH \overset{k''_2}{\rightleftarrows} HmNHCOO^- + H^+ \quad (6.12)$$

$$CO_2 + O_2HmNH_2 \underset{kb'_3}{\overset{kf'_3}{\rightleftarrows}} O_2HmNHCOOH \overset{k''_3}{\rightleftarrows} O2HmNHCOO^- + H^+ \quad (6.13)$$

$$O_2 + HmNH_2 \underset{kb'_4}{\overset{kf'_4}{\rightleftarrows}} O_2HmNH_2 \quad (6.14)$$

$$HmNH_\beta^+ \overset{k''_5}{\rightleftarrows} HmNH_2 + H^+ \quad (6.15)$$

$$O_2HmNH_3^+ \overset{k''_6}{\rightleftarrows} O_2HmNH_2 + H^+ \quad (6.16)$$

where Hm denotes a Heme group, $NHCOOH$ denotes a terminal amino group on the α and β chains to which carbon dioxide can bind and the terms kf'_i and kb'_i denote the forward and backward rate constants. The set of ODE's resulting from these reactions could be simplified to a system of algebraic equations at chemical equilibrium and the oxyhemoglobin and carbamino saturations of arterial blood could be derived:

$$S_a O_2 = \frac{K_{HbO_2}\, \alpha_{O_2}\, P_a O_2}{1 + K_{HbO_2}\, \alpha_{O_2}\, P_a O_2} \qquad (6.17)$$

$$S_a CO_2 = \frac{K_{HbCO_2}\, \alpha_{CO_2}\, P_a CO_2}{1 + K_{HbCO_2}\, \alpha_{CO_2}\, P_a CO_2} \qquad (6.18)$$

where $S_a O_2$ and $S_a CO_2$ are the oxyhemoglobin and carbamino saturations respectively, $P_a O_2$ and $P_a CO_2$ are the oxygen and carbon dioxide tensions in the arterial plasma, with solubility coefficients α_{O_2} and α_{CO_2} respectively. K_{HbO_2} and K_{HbCO_2} are parameters based on the complex interaction of hemoglobin with various other substances which take the form:

$$K_{HbO_2} = \frac{K_4'\left(K_3'\,\alpha_{CO_2}\, P_a CO_2\left(1 + \frac{K_3''}{C_a H^+}\right) + \left(1 + \frac{C_a H^+}{K_6''}\right)\right)}{K_2'\,\alpha_{CO_2}\, P_a CO_2\left(1 + \frac{K_2''}{C_a H^+}\right) + \left(1 + \frac{C_a H^+}{K_5''}\right)} \qquad (6.19)$$

$$K_{HbCO_2} = \frac{K_2'\left(1 + \frac{K_2''}{C_a H^+}\right) + K_3' K_4'\left(1 + \frac{K_3''}{C_a H^+}\right)\alpha_{O_2}\, P_{O_2}}{\left(1 + \frac{C_a H^+}{K_5''}\right) + K_4'\left(1 + \frac{C_a H^+}{K_6''}\right)\alpha_{O_2}\, P_{O_2}} \qquad (6.20)$$

where in this case the K_i' and K_i'' refer to equilibrium and ionization constants respectively and $C_a H^+$ denotes the hydrogen ion concentration in the arterial plasma. The equilibrium constant K_4' for the association of oxygen with hemoglobin reaction in (6.13) has a dependence on the oxygen tension, carbon dioxide tension, the concentration of hydrogen ions and DPG as well as the temperature of the blood (as outlined in Chapter 2) and is defined through the power law proportionality equation as:

$$K_4' = K_4'' \left(\frac{P_a O_2}{P_a O_{2sp}} \right)^{n_0} \left(\frac{C_a H^+}{C_a H_{sp}^+} \right)^{-n_1} \left(\frac{P_a CO_2}{P_a CO_{2sp}} \right)^{-n_2} \left(\frac{C_a DPG}{C_a DPG_{sp}} \right)^{-n_3} \left(\frac{T}{T_{sp}} \right)^{-n_4}$$

(6.21)

where $C_a DPG$ denotes the concentration of DPG in the blood, T is the blood temperature, the subscript SP retains its previous meaning of a normal or 'set point' value, and the exponents are empirically determined values described by the functions:

$$n_1 = -6.775 - 2.0372 \log \left(C_a H^+ \right) + 0.1235 \left(\log \left(C_a H^+ \right) \right)^2 \qquad (6.22)$$

$$n_2 = -0.008765 + 0.00086 \, P_a CO_2 + 6.3 \times 10^{-7} \, P_a CO_2^2 \qquad (6.23)$$

$$n_3 = 0.2583 + 26.6978 \, C_a DPG - 917.69 \, C_a DPG^2 \qquad (6.24)$$

$$n_4 = 1.6914 + 0.06186 \, T + 0.00048 \, T^2 \qquad (6.25)$$

The concentration of oxygen in the arterial blood depends upon the amount that is dissolved in the plasma, both inside and outside of the red blood cells and the amount which is bound to hemoglobin, and is defined as:

$$C_a O_2 = ((1 - Hct) \, W_{pl} + Hct \, Wrbc) \, \alpha_{O_2} \, P_a O_2 + 4 \, C_a Hb \, S_a O_2 \qquad (6.26)$$

where W_{pl} and W_{rbc} are the fractional water spaces of the plasma and red blood cells and Hct is the blood *hematocrit* (the ratio of the volume of red blood cells to a given volume of blood). The concentration of carbon dioxide in the arterial blood depends upon the amount that is dissolved in the plasma, the amount which has formed bicarbonate and the amount which is bound to hemoglobin, and is defined as:

$$C_a CO_2 = ((1 - Hct)\, W_{pl} + Hct\, Wrbc)\, \alpha_{CO_2}\, P_a CO_2$$
$$+ ((1 - Hct)\, W_{pl} + Hct\, Wrbc\, R_{rbc}) \left(\frac{K_1\, \alpha_{CO_2}\, P_a CO_2}{R_{rbc}\, CaH+} \right) \quad (6.27)$$
$$+ 4\, Hct\, C_a Hb\, S_a CO_2$$

where R_{rbc} is the *Gibbs-Donnan ratio* for electrochemical equilibrium across the red blood cell membrane, which is related to the concentrations of hydrogen and bicarbonate ions inside and outside of the red blood cell. The complete list of parameters required for this oxygen carbon dioxide concentration model is given in Table 6.1.

Table 6.1: Constant parameters used in the oxygen and carbon dioxide concentration model of Dash et al [35] and incorporated in the present study.

K_1	7.43×10^{-7} M	Hct	0.45
K_2''	1×10^{-6} M	W_{pl}	0.94
K_2'	29.5 M^{-1}	W_{rbc}	0.65
K_3''	1×10^{-6} M	W_{bl}	0.81
K_3'	25.1 M^{-1}	$P_a O_{2sp}$	100mmHg
K_4''	202123 M^{-1}	$P_a CO_{2sp}$	40mmHg
K_4'	202123 M^{-1}	$C_a H_{sp}^+$	57.5×10^{-9} M
K_5''	2.63×10^{-8} M	$C_a DPG_{sp}$	4.65×10^{-3} M
K_6''	1.91×10^{-8} M	T_{sp}	37° C
$C_a Hb$	5.18×10^{-3} M	R_{rbc}	0.69
α_{O_2}	1.46×10^{-6} M/mmHg	α_{CO_2}	3.27×10^{-5} M/mmHg

It was also decided that the chemical cascade involved in the metabolic pathways between oxygen, carbon dioxide and intracellular calcium levels need not be included as this again involves a level of complexity which is of no advantage and still not fully elucidated at present. Furthermore it was decided that the influence of oxygen on intracellular calcium would not be included, since it is known that a large reduction in oxygen delivery is required before the byproducts of anaerobic respiration will cause dilation, at which point the increase in carbon dioxide levels (due it its lack of removal by the blood) will have induced vasodilation. The new model was developed by beginning with a dynamic for the concentration of carbon dioxide in the tissue

C_tCO_2, similar to the model of Gutierrez[54]:

$$\frac{dC_tCO_2}{dt} = CMRO_2 - CBF\,(C_tCO_2 - C_aCO_2) \qquad (6.28)$$

where the arterial carbon dioxide concentration is calculated from Equation 6.27. This equation has the physiological interpretation that the difference between the rate of production of carbon dioxide in the tissue (equal to the consumption of oxygen, hence the incorporation of $CMRO_2$) and the rate of removal of carbon dioxide by the blood is equal to the rate of change of carbon dioxide in the tissue. Any further chemical reactions such as the reduction in pH or a production of nitric oxide are then ignored, but implicitly incorporated into the dynamic for CVR which is postulated to be a function of C_tCO_2 as:

$$\frac{dAut_{CVR}}{dt} = G_{CVR}\,(C_tCO_{2SP} - C_tCO_2) \qquad (6.29)$$

where Aut_{CVR} represents an autoregulation *activation* for CVR to change (for reasons which will be discussed shortly) and G_{CVR} is a proportional gain, which was chosen to be 7 in order to replicate the dynamics observed by Newell et al [112]. C_tCO_{2SP} is the normal or set point carbon dioxide concentration in the cerebral tissue under homeostatic equilibrium and is given as the steady state solution of (6.28) as:

$$C_tCO_{2SP} = C_aCO_2 + \frac{CMRO_2}{CBF_{SP}} \qquad (6.30)$$

Equation 6.29 has the physiological interpretation that when C_tCO_2 levels deviate from their normal value, there is an activation for smooth muscle to alter its tone and hence CVR. The reason for having the dynamics in (6.29) based on an activation and not CVR directly arises from the desire to incorporate a more physiologically accurate method of implementing the limits of autoregulation. The determination of the limits in the previous model involved a simple logical check during the evaluation of CVR for the next time step in the simulation. If the computed value of CVR was outside the limits, then either the upper or lower limit was returned respectively, resulting in a sharp transition from the autoregulated to the un-autoregulated range of $MABP$. The new method formulated allows instead for the activation of changes in CVR (i.e. pertaining to the concentrations of certain metabolites in the brain tissue or smooth muscle cell) to attain a value related only to metabolism and blood flow, whereas the actual CVR is limited by the phenomenological sigmoidal function:

$$CVR = \frac{CVR_{LL} + CVR_{UL}e^{Aut_{CVR}-C}}{1 + e^{Aut_{CVR}-C}} \tag{6.31}$$

where CVR_{UL} and CVR_{LL} are the upper and lower limits of autoregulation respectively. The sigmoidal function in (6.31), illustrated in Figure 6.4(a) has the property that regardless of the activation for alterations in smooth muscle tone, CVR cannot vary outside the upper and lower limits set in the function. The idea behind this implementation of the limits is that as the upper or lower limits are approached the ability of CVR to further increase or decrease approaches zero, leading to a more rounded off limit as illustrated in the autoregulation curve in Figure 6.4(b).

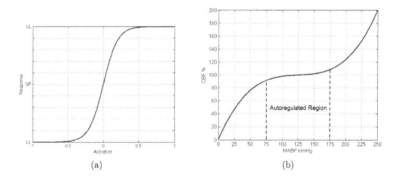

(a)   (b)

Figure 6.4: (a) A generic plot of the sigmoidal function incorporated in cerebrovascular resistance limits and (b) the effect on the autoregulation curve.

The constant C in (6.31) was incorporated to shift the sigmoidal curve along the x axis so that at zero activation, CVR was at its set point, for the reason being that without it, the value returned at zero activation would be midway between the upper and lower limits. As this may not be the case the constant was evaluated as:

$$C = -\ln\left(\frac{CVR_{SP} + CVR_{LL}}{CVR_{UL} + CVR_{SP}}\right) \tag{6.32}$$

so that no matter where the set point falls between the upper and lower limits, its value will be returned at zero activation. In keeping with the deviation from the autoregulation curve of Figure 6.2, it was decided to improve on the limits of autoregulation as well. In the experimental work of Strandgaard et al [142], volunteers were subjected to an infusion of *angiotensin II amide* to raise their $MABP$ and then subsequent infusions of *trimethaphan camsylate* combined with a $20-25degree$ head-up tilt manoeuver to lower $MABP$ in a series of steps. During the experiment, CBF was monitored so that the lower limit of autoregulation could be observed and furthermore, the lowest tolerable blood pressure before the evidence of ischaemia occurred was monitored. Of the uncontrolled hypertensive, controlled hypertensive and normotensive subgroups studied, it was discovered that the lower limit of autoregulation occurred at 79%, 72% and 74% of the resting $MABP$ respectively. Figure 6.5 is an example of part of the autoregulation curve obtained from one of the controlled hypertensive volunteers, illustrating the normal resting $MABP$ and where the lower limit of autoregulation was determined. As the volunteers used to construct the segmented circle of Willis models for the present study were all young healthy individuals, it was therefore decided to use the result from the normotensive group that the lower limit of autoregulation for these volunteers would occur at 74% of the normal resting $MABP$. Furthermore, since all of the simulations are designed at investigating factors that will lead to insufficient CBF and ischaemia, the upper limit of autoregulation is of less importance as it will not be approached. For this reason it was decided to simply apply a symmetric condition where the upper limit in (6.31) would be equal to CVR_{SP} plus the difference between the set point and the lower limit CVR.

Further improvement in the model came about from the inclusion of another physiological variable known as *oxygen extraction fraction (OEF)*, which is the proportion of the oxygen available in the arterial blood that is actually consumed to meet metabolic demands, and in most normal cases is around $30-40\%$. The reason for its inclusion in the model is that the crucial result desired from the simulations is oxygen delivery to the various cerebral territories supplied by the efferent arteries

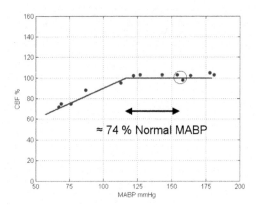

Figure 6.5: A sample result from the study of Strandgaard et al, illustrating lower half of an autoregulation curve from one volunteer of the hypertensive subgroup examined.

of the circle of Willis, defined as:

$$O_2 Delivery = C_a O_2 \cdot CBF \cdot OEF \qquad (6.33)$$

So while technically not a part of cerebral autoregulation, OEF plays an important role in maintaining oxygen delivery to the brain tissue as it in effect provides a second reserve when CBF cannot be maintained through vasodilation alone [41, 75]. To incorporate its effect a second dynamic was created:

$$\frac{dAut_{OEF}}{dt} = G_{OEF} \left(CMRO_2 - CBF \cdot C_a O_2 \cdot OEF \right) \qquad (6.34)$$

where Aut_{OEF} is an activation for oxygen extraction to change and G_{OEF} is a pro-portional gain, chosen to be 100. Equation 6.34 has the physiological interpretation that if the metabolic demand for oxygen is not met by the oxygen delivery, then there is a drive for OEF to increase, or conversely if the oxygen delivery exceeds the metabolic requirements then there will be a drive for it to decrease. It should be noted that while autoregulation represent an active control system, OEF is a passive phenomenon where oxygen diffuses into the tissue down a concentration gradient. In terms of incorporating the limits of oxygen extraction, the lower limit is intuitively zero, but would not be approached in simulations designed to predict ischaemia. Instead the upper limit of OEF is of much more importance and while it may be thought that the upper limit be 100% (i.e. all of the available oxygen in the blood is utilized) a number of studies have shown that ischaemia may result when OEF increases to above 77% [99, 135]. An upper limit of 77% was therefore specified for the present study and the actual value of OEF determined during the simulation calculated using the same sigmoidal function as for the autoregulation activation:

$$OEF = \frac{OEF_{LL} + OEF_{UL}e^{Aut_{OEF} - O}}{1 + e^{Aut_{OEF} - O}} \tag{6.35}$$

where the constant O is incorporated for the same reason that the set point for OEF will not necessarily be the exact mid point between the upper and lower limits and is evaluated as:

$$O = -\ln\left(\frac{OEF_{SP} + OEF_{LL}}{OEF_{UL} + OEF_{SP}}\right) \tag{6.36}$$

For incorporation with the discrete forms of the continuity and momentum equations during the simulation, the 3 ODE's were solved using a *Runge-Kutta* fourth order scheme. As the system of equations takes the form $\mathbf{y}' = f(\mathbf{y}, t)$ where:

$$\mathbf{y} = \begin{vmatrix} C_t CO_2 \\ Aut_{CVR} \\ Aut_{OEF} \end{vmatrix} \qquad f(\mathbf{y}, t) = \begin{vmatrix} CMRO_2 - CBF\left(C_t CO_2 - C_a CO_2\right) \\ G_{CVR}\left(C_t CO_{2SP} - C_t CO_2\right) \\ G_{OEF}\left(CMRO_2 - CBF \cdot C_a O_2 \cdot OEF\right) \end{vmatrix} \tag{6.37}$$

then the discrete values of $\mathbf{y}$ are computed using the relation:

$$\mathbf{y}^{t+1} = \mathbf{y}^t + \Delta t \sum_{i=1}^{4} b_i Y'_{ti} \tag{6.38}$$

where:

$$b_i = \begin{bmatrix} \frac{1}{6} & \frac{1}{3} & \frac{1}{3} & \frac{1}{6} \end{bmatrix} \tag{6.39}$$

and Y'_{ti} is defined as:

$$Y'_{ti} = f(Y_{ti}) \tag{6.40}$$

with the internal stages Y_{ti} given by:

$$Y_{ti} = \mathbf{y}^t + \Delta t \sum_{i=1}^{4} \alpha_{ij} Y'_{ti} \tag{6.41}$$

where:

$$\alpha_{ij} = \begin{bmatrix} 1 & \frac{1}{2} & \frac{1}{2} & 1 \end{bmatrix} \tag{6.42}$$

In contrast to the idealized autoregulation model, where the deviation in CBF was the driving cause for a change in CVR, the driving cause for the patient specific model is the deviation in tissue carbon dioxide levels from their set point value. While this set point, evaluated with (6.30), depends upon the normal level of CBF, it is also a function of the normal arterial carbon dioxide concentration and the metabolic rate. While the normal arterial carbon dioxide concentration is evaluated through (6.27) the incorporation of the metabolic rate poses a problem. The reason for this problem is that metabolic rate data is generally known on a per unit mass of brain tissue basis (i.e. mL/g/min), hence for unit consistency in (6.30), either the metabolic rate must be known for the cerebral territory supplied by a particular cerebral artery (i.e. mL/min), or the CBF must be converted to a per unit mass of brain tissue basis. Since no data was found which could provide metabolic rates for the various cerebral territories it was decided to use the average value for the mass of the human brain m_{TOT} of 1400g [136] and assume that amount of blood flow in a given cerebral artery is proportional to the mass of the cerebral territory that it perfuses with blood. Following from this assumption the mass of a given cerebral territory m_i can be defined:

$$m_i = m_{TOT} \frac{CBF_{SP,i}}{CBF_{SP,TOT}} \qquad (6.43)$$

where $CBF_{SP,i}$ is the set point level of CBF through a given cerebral artery and $CBF_{SP,TOT}$ is the total amount blood flow into the circle of Willis. By then specifying the set point flowrates for each artery on a per unit mass basis, Equation 6.30 would yield the same set point concentration of carbon dioxide for every cerebral tissue.

Unlike the idealized circle of Willis models, and in keeping with the desire of making the CFD simulations patient specific it was decided to measure the normal flowrates of the volunteers in vivo. The measurements were made with the incorporation of *Phase Contrast (PC)* MRI and the PC scans were performed on the five volunteers at the same time at which the TOF scan was performed to image their cerebral vasculature. While a detailed explanation of phase contrast MRI is given in Appendix A, the basic principal is that a reversal of one of the magnetic field gradients can cause a change in the received MR signal which is directly proportional to the velocity of the blood. Furthermore, with the use of *CINE* phase contrast MRI (where an *electrocardiogram* of the individual undergoing the scan is taken) the variation in CBF throughout the cardiac cycle can be obtained.

While the use of phase contrast MRI to quantitatively measure CBF in the internal carotid and basilar arteries is common, with a wide body of research providing measurements [24, 97], its use in measuring CBF in the efferent arteries of the circle of Willis is less common. One such study by Enzmann et al [43] however, did quantitatively measure CBF in the efferent arteries, as well as the afferents. The results (Table 6.2) show that the average total inflow into the circle of Willis postulated by Hillen corresponds reasonably well with the experimental data. Furthermore, the flowrates determined by the 6:3:4 resistance ratio are similar in the ACA's and only slightly overestimated in the MCA's. The major difference lies in the average flowrate through the PCA's which is overestimated by almost a factor of three. Furthermore the results show that there is a reasonable variation between individuals (illustrated by the standard error) and between the left and right cerebral arteries. The validity of these results may be limited by the small number of individuals studied and the resolution of the scans, but they are in keeping with the results of many other studies [88] showing that the amount of CBF does vary among the population. A further advantage of incorporating PC MRI into the present study is that the dataset generated would expand upon the limited information available regarding PC measurements in the efferent arteries, as well as provide some clarification as to how blood flow in the efferent arteries was affected by anatomical variations of the circle of Willis.

Table 6.2: CBF and diameter values in the major cerebral arteries measured by Enzmann et al [43].

Artery	Left			Right		
	Diameter	CBF_{mean}		Diameter	CBF_{mean}	
	mm	mL/min	mL/cycle	mm	mL/min	mL/cycle
ICA	5.29 ± 0.68	337 ± 19	5.1 ± 0.3	5.29 ± 0.68	302 ± 21	4.6 ± 0.4
BA	4.22 ± 0.43	161 ± 11	2.4 ± 0.2	–	–	–
ACA	3.39 ± 0.68	75 ± 10	1.1 ± 0.2	3.57 ± 0.65	88 ± 11	1.3 ± 0.2
MCA	4.06 ± 0.58	108 ± 7	1.7 ± 0.1	4.37 ± 0.55	127 ± 7	1.8 ± 0.2
PCA	3.39 ± 0.68	53 ± 4	0.8 ± 0.1	3.39 ± 0.43	51 ± 4	0.8 ± 0.1

The procedure for obtaining the 2D phase contrast images in the efferent and afferent arteries involves initially performing a TOF scan in order to provide a reference image for the *imaging planes* to be defined relative to. The planes scanning for all of the efferent arteries are subsequently defined using the standard MR acquisition software in order to interactively position the plane to be as normal as possible to the centreline of the artery of interest (Figure 6.6(a)). Since the cerebral arteries are not generally normal to any of the standard acquisition planes (i.e. transverse, sagittal or coronal), the position of the plane must be iteratively repositioned from other orthogonal views in order to ensure its normality to the artery, before the data can be acquired (Figures 6.6(b) and 6.6(c)). The scanning parameters used in the 2D phase contrast protocols are given in Table 6.3.

One parameters involved in a phase contrast scan is the *encoding velocity* (V_{ENC}), which is the maximum velocity measurable by the MRI scanner (see Appendix A). The choice of encoding velocity is of great importance in the acquisition of good quality data. Since the signal generated by the flowing blood depends upon its velocity relative to the encoding velocity, it should ideally be set as close as possible to the maximum blood flow velocity in the artery over the cardiac cycle. As the maximum blood flow velocity cannot be known in advance however, in keeping with good radiographic practices [93], the initial scans experimented with different values for V_{ENC} to find the minimum values in each artery that could be used before aliasing occurred. The results chosen for the remainder of the study were encoding velocities of 90cms^{-1} in the ICA's and BA, 80cms^{-1} in the MCA's and 70cms^{-1} in the ACA's and PCA's.

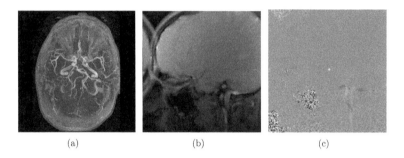

(a)   (b)   (c)

Figure 6.6: (a) A transverse reference MIP showing the placement of a scanning plane for determination of the blood flow through the LMCA, (b) the magnitude image of the PC scan normal to the scanning plane and (c) the phase image, illustrating the LMCA velocity as a white circle.

The result of the CINE PC MR scan is a series of images each corresponding to a particular point in the cardiac cycle, but the acquisition of the PC data is only part way to obtaining a measurement of the CBF in a given artery. While the velocity information is embedded in the phase images from the scan, the calculation of the flowrate from an image requires the specification of a region of interest around the lumen of a vessel. To this end an interactive program was developed to process the phase contrast scans and calculate the efferent flowrates (and for reasons that will be discussed in the next chapter, the afferent flowrates) of the segmented models. The DICOM data format of the 2D images uses 16 bit integers to represent the pixel intensities and with the protocol used by the GE 1.5T in the present study, the pixel velocity in cms^{-1} was one tenth the pixel intensity, meaning that a pixel intensity of say 500 would correspond to a velocity of 50cms^{-1}. Furthermore, since the direction of blood flow is embedded in the image the intensity range of the images was $-10\,V_{ENC}$ to $+10\,V_{ENC}$.

The procedure for obtaining the cerebral flowrates using the interactive software followed essentially the same procedure of manipulating the raw phase contrast data as was used on the time of flight data in order to create the segmented models. This procedure will therefore be described in the same manner, illustrating by example

Table 6.3: Phase Contrast parameters used in the acquisition of the MRI scans. Note: the definition of these parameters is given in Appendix A.

TR	7.65ms	Matrix Size	256×256
TE	3.16ms	Pixel Spacing	0.7mm
NEX	4	Slice Thickness	8mm
Flip Angle	20°	Points per Cycle	30

the various operations. The standard format of the PC MRI data is to have one phase image and one magnitude image for every point within the cardiac cycle. The general procedure for operating on the data would involve reading all of the phase and magnitude images into separate 3D arrays for all of the MRI datasets from each of the efferent and afferent scans. The major difference compared to the 3D TOF array is that rather than each image corresponding to a transverse location in the brain, it corresponds to a point in time. While there is no velocity information associated with the magnitude images, they can still be of use in that often they show more clearly where the edge of a given artery is compared to a phase image and in fact a number of studies using PC MRI to measure flowrates [43] use the magnitude images to locate the ROI, then map it back onto the phase images. This option was hence included in the software in order to provide the best estimate of the lumen of an artery. Since the internal carotid and basilar arteries traverse vertically into the circle of Willis essentially in parallel then all three arteries could be included in a single scan (Figures 6.7(a) and 6.7(b)), minimizing the amount of time required. Furthermore the anterior cerebral arteries course essentially parallel to each other, so both of these arteries could be acquired in one scan. For the remaining middle and posterior cerebral arteries however, a separate scanning plane was required for each artery, meaning that a total of six scans was needed to acquire all of the afferent and efferent boundary condition information.

Similar to the manipulations of the TOF data, the process would involves a rectangular ROI around the artery or arteries of interest (Figure 6.8(a)) and reducing the dataset (Figure 6.8(b)) to remove unwanted data and speed up subsequent operations.

The major difficulty in the determination of the flowrates is the placement of the ROI around the lumen of an artery, where different choices have the potential to produce very different calculations of the flowrate. The difficulty arises due to the limited resolution of the scan (Figure 6.9(b)) where typically there are only

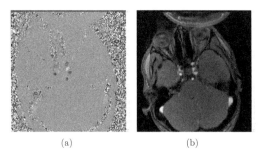

Figure 6.7: A transverse scanning plane acquiring velocity data through the internal carotid and basilar arteries, showing (a) the phase image and (b) the magnitude image.

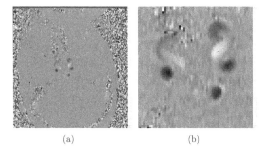

Figure 6.8: (a) A rectangular ROI drawn around the arteries of interest in a given PC dataset and (b) the reduced dataset.

between $5-10$ pixels across the diameter. As was mentioned previously, the phase images give the direction of flow as well as its magnitude, and this means that the intensity scale range is from $\pm 10 V_{ENC}$, where bright pixels correspond to flow in one direction, dark pixels to flow in the opposite direction, and zero velocity is in the mid range (i.e. grey). Typically, the region surrounding the lumen is in the mid range with small fluctuations around zero (Figure 6.10(a)), meaning that if the ROI is drawn too far away from the lumen then the contributions to the overall flowrate in the ROI from the surrounding tissue may be negative and result in an underestimation of the flowrate through the artery. If the absolute value of the pixel intensity was used instead in the calculation then the result would be an overestimation of the flowrate through the artery. What is needed for an optimal calculation is to locate the closest region around the lumen where the pixel

velocities become negative and to selectively remove them from them from the data
(meaning that they make no contribution to the flowrate if included in the ROI.
To improve on the flowrate measurements through each artery the reduced datasets
(Figure 6.9(a)) would initially have the colourmaps switched to RGB to enhance the
contrast between different signal intensities (Figure 6.9(b)) and then be interpolated
so that a more obviously circular cross section could be visualized (Figure 6.9(c)).
As is illustrated in Figures 6.10(a) and 6.10(b), interpolating the dataset has a
negligible effect on the velocity profile.

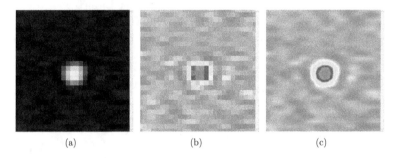

Figure 6.9: The reduced PC dataset of an example LMCA illustrating (a) the greyscale
colourmap, (b) and RGB colourmap and c) the interpolated image, giving a much clear
visualization of the velocity profile.

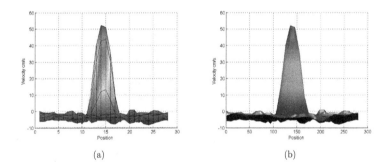

Figure 6.10: A surface plot of the LMCA illustrating (a) the coarse waveform and (b) the
interpolated waveform. The key point of the comparison is that firstly the interpolation of
the data has negligible effect on the velocity profile and b) the region of negative velocities
surrounding the lumen of each artery.

Following the interpolation the data in the region around the lumen can be cleared (Figure 6.11(a)) so that a *zero level*, or intensity corresponding to a velocity of zero, can be clearly seen. By using the eraser tool, any pixels around the lumen which are darker than the zero level (meaning that they correspond to negative velocities) can be selectively removed. As the darkest (most negative) pixels are deleted from the dataset the zero level intensity will change accordingly, approaching the darker end of the colourmap (Figure 6.11(b)) until there are no pixels surrounding the lumen which are darker than the zero level. At this point a ROI may be drawn around the lumen with the assurance of no negative contributions from the surrounding tissue. Typically the resulting flowrates would be approximately 1-5% higher using this procedure compared to directly drawing the ROI around the raw data. Since it was shown in Figures 6.10(a) and 6.10(b) that the changes in velocity profile were negligible it is therefore assumed that this difference in calculated flowrate is the result of the removal of any negative contributions from the surrounding tissue and better choice of the ROI around an arterial lumen.

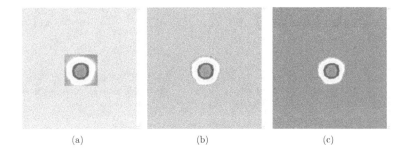

(a)   (b)   (c)

Figure 6.11: The reduced PC dataset of an example LMCA illustrating (a) the data cleared outside of a rectangular ROI around the lumen with the zero level visible as light orange, (b) the zero level becoming a darker orange as the negative pixels surrounding the lumen are removed and (c) the isolated velocity profile.

With the best possible estimation of the lumen of an artery, the flowrate for a given artery at a certain point in the cardiac cycle can be obtained using the relation:

$$CBF = \int_{ROI} \mathbf{u} \cdot d\mathbf{A} \approx \sum_{ROI} u_{PIX} A_{PIX} \tag{6.44}$$

where the approximation is used for the discrete data given in the PC datasets. A_{PIX} is the pixel area and is the product of the pixel spacing in the x and y directions (Table 6.3), set by the MRI scanner and obtainable in the DICOM information associated with each image. In order to determine how the flowrates vary as a function of time, the time period of the cardiac cycle T and the timestepsize between the discrete points measured in the cardiac cycle Δt were calculated as:

$$T = \frac{60}{HR} \qquad \Delta t = \frac{60}{HR \cdot N} \tag{6.45}$$

where HR is the heart rate in beats per minute (bpm) and N is the number of points in the cardiac cycle. By applying (6.44) to every slice in the 3D dataset, for every artery, the flowrates over the cardiac cycle as a function of time could be calculated (Figure 7.2(a)). For the efferent boundaries it is the determination of the set point flowrates which is of interest in order to determine the reference $C_t CO_2$ levels. The set point flowrates were taken to be the average flowrate over the cardiac cycle, evaluated as:

$$CBF_{SP} = \frac{1}{T} \int_0^T CBF \, dt \approx \frac{1}{T} \sum_{n=1}^{N} \frac{CBF^n + CBF^{n-1}}{2} \Delta t \tag{6.46}$$

using a trapezoidal numerical integration method as an approximation to the analytical integral.

One important result of using patient specific cerebral flowrates is that since they are known to vary between individuals, and since the metabolic rate has been assumed for every circle of Willis model, then the calculated concentrations of carbon dioxide (evaluated via Equation 6.30) in the cerebral tissue will vary between individuals. Furthermore, since the concentration of oxygen in the arterial blood is assumed to be the same for every circle of Willis model, then the oxygen extraction fractions will also vary between individuals. For the present study practical limitations excluded these measurements for the five individuals undergoing the MRI scans, so this result will have to be validated in future work. While it is known that OEF does vary among the population [99, 135] and it is probably quite likely that C_tCO_2 varies also, the numerical values resulting from the patient specific autoregulation model cannot be proven as correct. It does seem reasonable however, that the amount of oxygen required per unit mass of cerebral tissue be kept as the constant between all models, and since the cerebral flowrates were measured as varying between individuals, one can accept the resulting differences in tissue carbon dioxide levels and oxygen extraction fractions.

Chapter 7

Fluid Dynamics Modelling

7.1 Introduction

With the generation of the circle of Willis geometries and the autoregulation models now defined, the purpose of the present chapter is to outline how the governing fluid dynamic equations are solved on the 3D grids and furthermore how the autoregulation models are coupled into the simulation. This chapter will begin by addressing the method for the solution of the governing equations, continue with two important issues relating to physiological hemodynamic simulations, namely the blood viscosity model and the boundary conditions, and end with an outline of how the simulations were performed.

7.2 Governing Equations and Discretization

The mathematical considerations behind the derivation of the governing equations of fluid mechanics from first principals and an explanation of their numerical solution method is an extremely involved procedure. This chapter will not attempt to cover all of this information, but merely state the final form of the governing equations and subsequently their discrete approximations, referring to Appendix C where the complete derivation is given. Given that blood is an incompressible fluid, the two physical principals required to completely describe the flow field within the lumen of the circle of Willis are the *conservation of mass* and the *conservation of momentum*. Utilizing a *control volume* approach to apply these physical principals, the conservation of mass equation takes the form:

$$\int_{CS} \rho\,\mathbf{u} \cdot d\mathbf{A} = 0 \tag{7.1}$$

where ρ is the blood density, $\mathbf{u}$ is the 3D velocity vector field and CS is the bounding surface of the control volume CV. The conservation of momentum equation takes the form:

$$\frac{\partial}{\partial t} \int_{CV} \rho\,\mathbf{u}\,dV + \int_{CS} \rho\,\mathbf{u}\,\mathbf{u} \cdot d\mathbf{A} = \int_{CS} \eta\,\nabla \mathbf{u} \cdot d\mathbf{A} - \int_{CS} p\,I \cdot d\mathbf{A} \tag{7.2}$$

where η is the blood viscosity, p is the pressure field and A and V are the elemental area and volume of CV. In order to couple in the effects of alterations in CVR with the governing equations, an extra body force term was added into the momentum equation, such that when it is applied within the fluid domain of the porous blocks it takes the form:

$$\frac{\partial}{\partial t} \int_{CV} \rho\,\mathbf{u}\,dV + \int_{CS} \rho\,\mathbf{u}\,\mathbf{u} \cdot d\mathbf{A} = \int_{CS} \eta\,\nabla \mathbf{u} \cdot d\mathbf{A} - \int_{CS} p\,I \cdot d\mathbf{A} - \int_{CV} \eta\,CVR\,\mathbf{u}\,dV \tag{7.3}$$

where the extra body force term is essentially *Darcy's law* for porous media, assuming laminar flow [33]. The addition of this extra body force term has the effect of 'draining' momentum out of the flow, causing a large pressure gradient through the porous block. The loss in blood pressure is similar to that which would occur if the blood were flowing through the entire cerebrovascular network. Then, as the autoregulation mechanism causes CVR to alter and the body force term either increases or decrease, the pressure gradient through the porous block will alter, in a similar manner to that which would occur if the arteries and arterioles in the brain alter their diameter.

Given the complexity of the circle of Willis geometries, generation of a 3D computational grids must be achieved using *unstructured* meshing techniques, such as the TGrid scheme used on the idealized circle of Willis models and the adaptive cartesian method used on the patient specific models. The use of unstructured meshing places restrictions on the techniques available for the solution of the governing equations and for the present study the *finite volume method (FVM)* was chosen. The basic principal underlying the finite volume method is that the conservation equations (7.1) and (7.3) are applied to every cell in the computational grid, with the integrals replaced by discrete summations and the derivatives replaced by finite differences. In this formulation, the continuous pressure and velocity fields are replaced by a single value, stored at the centroid of the cell. The discrete form of (7.3) when applied to a given cell hence takes the form:

$$
\frac{3\rho\mathbf{u}_c^{t+1} - 4\rho\mathbf{u}_c^{t} + \rho\mathbf{u}_c^{t-1}}{2\Delta t} V_c + \sum_f^{N_{facenb}} \rho\,\mathbf{u}_f^{t+1}\,\mathbf{u}_f^{t+1} \cdot \mathbf{A}_f
$$
$$
= \sum_f^{N_{facenb}} \eta_f \nabla\mathbf{u}_f^{t+1} \cdot \mathbf{A}_f - \sum_f^{N_{facenb}} p_f \mathbf{A}_f - \eta_c\,CVR\,\mathbf{u}_c^{t+1}\,V_c
$$

(7.4)

where the superscript t refers to a value at the current timestep, to which the timestep size is Δt, the subscript f refers to a face of the control volume (implying that the values are evaluated at the centroid of a cell face, not the cell centroid itself), the total number of which comprising a cell is N_{facenb}, and the subscript c refers to a value at the centroid of the given cell. It should be noted that the face velocity, viscosity and area vectors $\mathbf{u}_f$, η_f and $\mathbf{A}_f$ respectively, are not expressed in terms of the given cell c for the reason that a face is shared by two cells. Following the discretization procedure, Equation 7.4 now forms a nonlinear algebraic equation. Furthermore, since the face values of $\mathbf{u}$ and η will be interpolated in (7.4) using the value stored at the centre of the cell c and the value at the centroid of the neighbouring cell which shares the given face f, the application of (7.4) to all of the cells in the computational grid, yields a coupled system of nonlinear algebraic equations. In order to proceed with obtaining a solution Equation 7.4 is *linearized* for a given cell c to take the form:

$$
a_c\mathbf{u}_c = \sum_n^{N_{cellnb}} a_n\mathbf{u}_n + b_c
$$

(7.5)

where the a_c and a_n are coefficients of the cell c and its neighbours n (the total number of which is N_{cellnb}), incorporating parts of the first three terms in (7.4) and b_c is a *source* vector, incorporating parts of the first term, as well as the fourth and fifth terms in (7.4). The overall system of equations can then be expressed in the linear form:

$$A\mathbf{u} = \mathbf{b} \tag{7.6}$$

where A is a diagonally dominant sparse matrix, and $\mathbf{b}$ is a column vector containing the known quantities. Although this resulting system of equations is linear and A could be inverted to provide the solution vector $\mathbf{u}$, the amount of memory required to invert a matrix is usually computationally prohibitive for practical problems. Furthermore it is an inefficient use of computer memory to store all the entries in a sparse matrix when most of the entries, except those centered about the main diagonal are zero. As a result, iterative methods are used to solve (7.6) and the method used for the present study is a *point-implicit Gauss-Seidel* method. It is important to note that (7.4) is a vector equation, with three components. Choosing a *segregated implicit* solution method, as was done for the present study, the three velocity components are each solved separately as separate matrix equations.

Thus far, no mention has been made of the discretization of conservation of mass equation, for the reason that its incorporation into the discrete system of equations is more complicated than that of the momentum equations. The solution of (7.6) at every iteration provides the updated velocity field u, but in order to provide this update the pressure gradient term embedded in 7.6 must be known. In the present study (as with most CFD problems) the pressure field is not known in advance, but arises as part of the overall solution. Since the conservation of momentum equations have been used to solve for the velocity field, the remaining equation, namely the conservation of mass equation, must be converted into an equation for pressure, such that the pressure field can be updated at every iteration and used in the solution of the velocity field. While the details of the conversion of the conservation of mass equation are given in Appendix C, this solution method (known as a *pressure based solution method*) involves obtaining an equation for *pressure correction* p' rather than pressure itself which, similar to (7.5), can be linearized to take the form:

$$q_c p'_c = \sum_{n}^{N_{neighbours}} q_n p'_n + r_c \tag{7.7}$$

where the q_c and q_n coefficients of the cell c and its neighbours n and r_c is a *source* vector involving pressure and velocity terms. Similar to (7.6), the system of equations resulting from the application of (7.7) can also be expressed in a matrix form and solved using the point-implicit Gauss-Seidel method. The overall solution procedure essentially involves initializing both the pressure and velocity fields at the centroids of every cell, and then iteratively solving the momentum equations to obtain an updated velocity field, using the updated velocity field to solve for the pressure correction, using the pressure correction to update the pressure field which is then used in the momentum equation to solve for the next updated velocity field, and repeating the process until the solution converges.

One issue relating to the solution of the governing equations is that of turbulence modelling. Theoretically the conservation of mass and momentum equations encompass all of the physics required to completely describe fluid flow, including its turbulent nature. Turbulence however, can occur on scales as small as the micron and millisecond spatial and temporal scales respectively, and when the governing equations are discretized over the spatial and temporal domains, the spatial grids and the timestep sizes are usually too large to capture turbulent phenomena. When a particular flow of interest is known to be turbulent, *turbulence models* may be incorporated as extra equations to be solved with the conservation of mass and momentum equations. It is known however that blood flow is laminar throughout most regions of the body with the exception of the aortic arch and occasionally in regions around stenosed blood vessels. Throughout the circle of Willis, the Reynolds numbers are typically of the order of 100 - 200, which is well below the laminar to turbulence transition region for flow in what is very nearly flow in a straight pipe. For this reason the flow has been assumed laminar and no turbulence models have been implemented for the present study, thereby leaving the investigation of whether or not turbulence is generated throughout the stenoses in the patient specific geometries for future work.

7.3   Blood Viscosity Models

One of the features of blood flow which complicates the overall solution of the governing equations is the well known fact that blood is a *non-Newtonian* fluid. As was outlined in Chapter 2, blood is composed mainly of plasma and red blood cells. The plasma is mainly water, containing some dissolved proteins and ions and exhibits a Newtonian nature. The non Newtonian effects of the blood are due to the red blood cells. As illustrated in Figure 7.1, blood undergoes shear thinning at shear rates greater than 100 s^{-1} due to the deformation of the red blood cells whereby the viscosity decreases and approaches an *infinite shear* viscosity. As the shear rate decreases the viscoelastic nature of blood becomes important and the viscosity increases due to the aggregation of the red blood cells approaching a *zero shear* viscosity. Incorporation of the non-Newtonian nature of blood into the momentum equation hence requires a more complicated viscosity model, rather than assuming a constant value, as would be the case for a Newtonian fluid approximation.

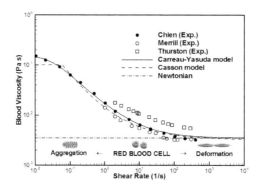

Figure 7.1: The Non Newtonian Effects of Blood [80]

All viscosity models are phenomenological in nature as they relate the viscous stresses to the rate of strain by fitting experimental data to proposed mathematical functions. Before explicitly stating the mathematical functions however, some important quantities in tensor algebra need to be introduced. It is a general rule of tensor algebra that for every symmetric tensor it is possible to determine three mutually orthogonal axes, known as *principal axes*. When the rate of strain tensor is expressed within this system of coordinates it contains only diagonal terms, known as *principal strains*, with all other non diagonal terms being equal to zero. This does not mean that there exist no shear strains on other planes, or that the element remains undistorted, only that in terms of the principal axes the fluid element is strained in three mutually perpendicular directions. Determining the principal axes and principal strains involves solving the eigenvalue problem:

$$|\lambda I - \bar{\epsilon}| = 0 \qquad (7.8)$$

where ϵ is the *rate of strain tensor* (see Appendix C). The principal strains are given by the eigenvalues λ_i and the principal axes are given by the eigenvectors. Evaluating the determinant yields the characteristic polynomial:

$$\lambda^3 - I_1\lambda^2 - I_2\lambda - I_3 = 0 \qquad (7.9)$$

where the scalar coefficients of the characteristic polynomial are important physical quantities known as *invariants* of the rate of strain tensor and are given as:

$$
\begin{aligned}
I_1 &= \epsilon_{xx} + \epsilon_{yy} + \epsilon_{zz} & (7.10)\\
I_2 &= \epsilon_{xy}^2 + \epsilon_{yz}^2 + \epsilon_{zx}^2 - \epsilon_{xx}\epsilon_{yy} - \epsilon_{yy}\epsilon_{zz} - \epsilon_{zz}\epsilon_{xx} & (7.11)\\
I_3 &= \epsilon_{xx}\epsilon_{yy}\epsilon_{zz} + 2\epsilon_{xy}\epsilon_{yz}\epsilon_{zx} - \epsilon_{xx}\epsilon_{yz}^2 - \epsilon_{yy}\epsilon_{zx}^2 - \epsilon_{zz}\epsilon_{xy}^2 & (7.12)
\end{aligned}
$$

Generally speaking, viscosity models for non Newtonian fluids propose that the viscosity is a function of the strain rate magnitude $\dot{\gamma}$:

$$\eta = f(\dot{\gamma}) \qquad (7.13)$$

where the strain rate magnitude is a function of the second invariant of the rate of strain tensor:

$$\dot{\gamma} = 2\sqrt{I_2} \qquad (7.14)$$

The reason for involving the second invariant is that non Newtonian fluids have a complex, non linear behavior and it is convenient to therefore make the model a function of the second invariant, which is itself composed of complex non linear terms.

It can be observed in Figure 7.1 that a Newtonian fluid assumption would be valid for shear rates greater than $100s^{-1}$, as would be the case in most of the major arteries, where the Newtonian viscosity may be set to the infinite shear viscosity. In smaller blood vessels such as the communicating arteries however, or large blood vessels where the blood flow has been reduced by a stenosis or occlusion, the shear rates may be considerably lower and the blood viscosity hence considerably higher. In such a case the assumption of a Newtonian fluid may no longer be valid. The viscosity model chosen for the present study is the *Carreau-Yasuda* model, given as:

$$\eta(\dot{\gamma}) = \eta_\infty + (\eta_0 - \eta_\infty)\left(1 + (\lambda\dot{\gamma})^a\right)^{\frac{n-1}{a}} \tag{7.15}$$

where η_0 is the zero shear viscosity and η_∞ the infinite shear viscosity, taken as 0.022Pa.s and 0.0022Pa.s respectively. The remaining constitutive parameters λ, a, n are taken as 0.11s, 0.644 and 0.392 respectively [52, 53]. As can be observed in Figure 7.1 this model more accurately predicts the variation in blood viscosity, especially at lower shear rates and is the model chosen for the present study.

7.4   Boundary Conditions

Another feature of the blood flow simulations which must be addressed is the choice of boundary conditions for the 3D geometries, since the specification of physiologically accurate boundary conditions is of crucial importance if the simulation results are to have any merit. Essentially the whole process of solving the governing equations can be thought of as propagating the flow conditions specified at the boundary into the fluid domain. The system of equations used for the present study form what is known as a mixed *parabolic - elliptic* system of equations, meaning that in order to obtain a solution, boundary conditions must be specified over the *entire* boundary of the fluid domain. Two major classes of boundary conditions exist, known as *Dirichlet* and *Neumann* conditions. Using Dirichlet conditions the value of the flow variable (either **u** or p) is specified on the boundary, whereas with Neumann conditions the derivatives of the flow variable is specified on the boundary. For reasons that will become apparent shortly, all of the boundary conditions used in the present study are of the Dirichlet type.

With the afferent and efferent terminations of the model discussed previously, the resulting boundaries requiring specification are the arterial wall around the circle of Willis, the afferent inlets at the level of the carotid and basilar (or vertebral) arteries, and the efferent outlets of the proximal branches of the major cerebral arteries. Considering first the arterial wall, it was decided to treat it as a rigid structure and impose the no slip condition. Despite the fact that the larger arteries closer to the heart such as the aorta and common carotid arteries do show an appreciable distensibility during the cardiac cycle [69, 138, 139], it is assumed that further up in the skull these effects are negligible. The justification for this assumption is that there is experimental evidence to show that even in the carotid bifurcation the change in diameter is small [79, 141] and that these changes in diameter have little effect on the blood flow [79]. The arteries comprising the circle of Willis are located distal to the carotid bifurcation and are further constrained from distending by the cerebral tissue fixed within the volume of the cranial cavity. While to the author's knowledge there has been no direct experimental research at present investigating the distensibility of the arteries comprising the circle of Willis and its effect on cerebral blood flow, it is thought that rigid walls are a valid assumption. Furthermore, the treatment of the arterial wall as a rigid structure removes the significant complication that would be added to the numerical solution of the governing equations if the arterial walls were able to distend.

For the remaining boundaries the options are to specify either pressure or velocity at the afferent boundary faces and either pressure, velocity, or the velocity gradient at the efferent boundary faces. For physiologically accurate simulations of cerebral blood flow one of these properties at that particular point in the cerebrovascular system must be known. The specification of velocity or flowrate (indirectly a specification of velocity) at the inlets has the advantage that these values may be measured experimentally in vivo, and similarly with the outlets (although the two cannot be used together). Instead velocity or flowrate inlet boundary conditions are commonly used in combination with velocity gradient or 'outflow' boundary conditions at the outlets. The only limitation of this choice is that it the flowrate through various parts of the model becomes fixed and will not be affected by changes in arterial blood pressure. In order to simulate how blood flow will be affected by changes in blood pressure resulting from occlusion or stenosis, the blood pressure itself must be the specified at both the inlets and outlets. While the inlet pressure can be assumed reasonably well to be equal to the overall systemic blood pressure, the physiological values for arterial blood pressures distal to the circle of Willis is unknown, posing a problem with this choice of boundary condition. Furthermore, relatively small changes in pressure gradient between the model inlets and outlets can cause significantly large and unphysiological changes in blood flow. So it is accurate values of cerebral blood flow which are required in the simulations, but it is the values of arterial blood pressure which must be specified at the model inlets and outlets. In order to circumvent this problem, and while simulating the physiology of the cardiovascular system, arterial pressures are specified at the model inlets and the downstream outlets of the porous blocks. The correct amount of blood flow through an artery is then achieved by the alteration of the body force term in the porous blocks, essentially altering the cerebrovascular resistance.

For the idealized models it was decided to implement a non-pulsatile mean arterial blood pressure since the resistance dynamic incorporated in the autoregulation model of this study would filter out the higher frequency oscillations anyway. The MABP for a healthy individual of 100mmHg was specified on all of the afferent faces and a normal venous pressure of 5mmHg on the efferent faces of the downstream ends of the porous blocks. For the patient specific models it was decided to investigate the effects of an afferent blood pressure variation over the cardiac cycle on the communicating flowrates and on the response of the resistance dynamic implemented in the autoregulation model. In order to generate the pressure waveforms on the afferent arterial inlets, the software developed for measuring the efferent flowrates was extended such that the measurement of the afferent flowrates waveforms over the cardiac cycle could be *mapped* to a pressure waveform. Since only 30 points were sampled during the cardiac cycle, the resulting transient flowrates had a somewhat jagged nature (Figure 7.2(a)). In order to produce a smoother continuous waveform, the flowrates could subsequently have *interpolating* and *smoothing* algorithms applied to them in much the same manner as the PC datasets, except that the algorithms used operate on 1D as opposed to 3D data (Figure 7.2(b)).

The pressure waveforms were then achieved by initially linearly mapping the

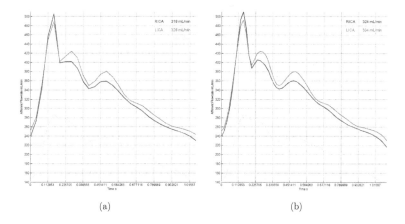

(a)   (b)

Figure 7.2: Transient flowrates through the three afferent arteries showing (a) the plot
of relatively jagged waveforms resulting from the 30 points sampled in the cardiac cycle
and (b) the plot of the smoothed waveforms. The mean flowrates over the cardiac cycle
are shown in the top right hand corner of each plot.

discrete transient flowrate waveform $CBF(n)$ to a pressure waveform $ABP(n)$ as:

$$ABP(n) = \frac{ABP_{SYS} - ABP_{DIA}}{CBF_{SYS} - CBF_{DIA}} CBF(n) + ABP_{DIA} \qquad (7.16)$$

where CBF_{SYS} and CBF_{DIA} are the systolic and diastolic (i.e. maximum and minimum) flowrates over the cardiac cycle and ABP_{SYS} and ABP_{DIA} the corresponding
systolic and diastolic blood pressures, chosen at present to be the standard systemic values for a healthy individual of 120 over 80mmHg. The assumption made
in doing so is that despite the variations in CBF through the afferent arteries, the
pressure losses along the arterial network up to the afferent terminations, where the
segmented models began, were equal (Figure 7.3(a)). Next the pressure waveforms
were subjected to a *Fourier decomposition* of the discrete data:

$$ABP(n) = \sum_{k=1}^{K} A_k \cos\left(\frac{2\pi(k-1)n}{N}\right) + \sum_{k=1}^{K} B_k \sin\left(\frac{2\pi(k-1)n}{N}\right) \qquad (7.17)$$

where K is the total number of A_k and B_k Fourier coefficients, evaluated as:

$$A_k = \sum_{k=1}^{\frac{N}{2}} \sum_{n=1}^{N} \frac{2}{N} ABP\left(n\right) \cos\left(\frac{2\pi\left(k-1\right)n}{N}\right) \tag{7.18}$$

$$B_k = \sum_{k=1}^{\frac{N}{2}} \sum_{n=1}^{N} \frac{2}{N} ABP\left(n\right) \sin\left(\frac{2\pi\left(k-1\right)n}{N}\right) \tag{7.19}$$

with:

$$A_{\frac{N}{2}} \leftarrow \tfrac{1}{2} A_{\frac{N}{2}} \qquad B_{\frac{N}{2}} \leftarrow 0 \tag{7.20}$$

With the Fourier coefficients calculated for each artery they could then be used during the fluid simulations to reconstruct the pressure waveforms as a function of time t:

$$ABP\left(t\right) = \sum_{k=1}^{K} A_k \cos\left(\frac{2\pi\left(k-1\right)t}{T}\right) + \sum_{k=1}^{K} B_k \sin\left(\frac{2\pi\left(k-1\right)t}{T}\right) \tag{7.21}$$

Because of the different lengths and diameters of the common carotid and vertebral arteries, it is likely that the pressure loss along each of these arteries will differ slightly and hence the assumption of equal systolic and diastolic pressures at the beginning of the model is not completely accurate. Since there is essentially no information available regarding the variations in ABP between the internal carotid and basilar arteries, this method remains the best estimate of a pressure waveform at that particular location but is a limitation of the current study which may investigated in future work. Preliminary results on the idealized models showed that there is very little pressure gradient between the ends of each communicating artery, which when combined with their small diameter results in very little flow though them under normal conditions. It requires a reasonable reduction in afferent APB for flow to be rerouted through them to other parts of the circle of Willis. Previous research [46] showed that blood flow through the communicating arteries may vary over the cardiac cycle (perhaps even reversing direction) due to the variation in transient ABP at the ends of these arteries over the cardiac cycle. Even though the pressure mappings used here assume the same systolic and diastolic flowrates, the shapes of the waveforms differ (Figure 7.3(a)), meaning that the pressure gradients across the communicating arteries will vary over the cardiac cycle and this result can be used to investigate whether or not significantly more blood flow is rerouted through them under normal conditions compared to the idealized models. It is well known from the Womersley solution of pulsatile flow in a straight rigid tube [169] that due to the inertia of the fluid, the pressure and flow will be out of phase. This result is illustrated in the sample simulation data of Figure 7.3(b), where the transient flowrate waveforms through the left efferent cerebral arteries of a segmented circle of Willis model are compared to the results of a simulation and can be seen to be shifted along the time axis. This result also gives confidence to the pressure mapped afferent boundary conditions in that the autoregulation mechanism incorporated in this particular simulation was altering the CVR in order to achieve the set point C_tCO_2 levels (and thereby set point or cardiac averaged CBF levels). The systolic and diastolic flowrates resulting from the simulation are determined entirely by the pressure waveforms, which correspond reasonably well. The phase shift of the numerical flowrates compared to the MRI data are of no physiological consequence and do not therefore introduce any error into the simulation. The difference in systolic flowrate between the MRI and simulation data is most likely due to too large a choice of timestep size, meaning that the systolic flowrates occurred in between the discrete points reconstructed for the simulation.

For the efferent pressures applied to the patient specific models, it was decided to continue to incorporate a constant pressure, but to increase the value to 30mmHg, as opposed to 5mmHg for the idealized models. The reason for doing so stemmed from the fact that when the CVR was at its set point (and the CBF in a given efferent artery was also at its set point) then the pressure on the upstream face of the porous block was only approximately 2mmHg lower than the pressure specified on the afferent inlets. As a result the pressure loss through the porous block was approximately 93mmHg. There was concern that given the relatively short length of the porous blocks, the large body force term required to generate this pressure drop may introduce some error into the numerical solution. While there were no observable effects in the idealized model results, it was decided to extend the length of the porous blocks and alter the definition of the efferent pressure that it repre-

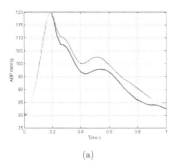
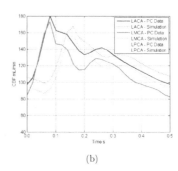

(a)   (b)

Figure 7.3: (a) An example of the pressure waveforms in the three afferent arteries and (b) the comparison between the efferent flowrate waveforms calculated from MRI to that produced by numerical simulation.

sented the pressure at the level of the capillary beds in the cerebrovascular network. While this would not effect the validity of the simulations, it would mean that CVR at the set point would be lower, as would the body force term in the porous blocks.

7.5   Simulation Procedure

The numerical values of CVR_{SP} required to achieve CBF_{SP} in the efferent arteries are not known a priori to performing the simulation, and must instead be determined as part of the overall solution. Furthermore, the same is true of the lower limits of autoregulation CVR_{LL}. The overall simulation procedure (Figure 7.4) would therefore begin with an initial guess for CVR (to which initial results showed the value of 1000mm^{-2} to be a close approximation) and determine these values for every efferent artery before moving on to simulating the effects of stenosis or occlusion. Using the result from the work of Strandgaard et al [142], that the lower limit of autoregulation was reached when the $MABP$ was lowered to 74% of its normal value, the simulations were begun with the afferent pressure inlets set at 75% of their $MABP$, allowing the autoregulation mechanism a period of approximately 20s simulation time, to alter CVR in all of the porous blocks until the value of C_tCO_{2SP} was achieved. At this point the current value of CVR in each porous block was recorded and assigned as CVR_{LL} for that artery. The $MABP$ was then restored to 100% (which would cause CBF_{SP} to increase above its set point and hence

C_tCO_2 below its set point) and the autoregulation mechanism was allowed another 20s of simulation time to alter CVR in all of the porous blocks until C_tCO_{2SP} was achieved. At this point the current value of CVR in each porous block was recorded and assigned as CVR_{SP} for that artery. Furthermore the value of CVR_{UL} could also be determined. At this point all of the information required for the autoregulation models was determined and the simulation of stenosis or occlusion could be performed.

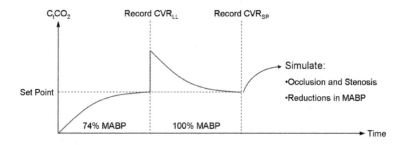

Figure 7.4: The simulation procedure for performing the circle of Willis CFD simulations.

For the idealized simulations the approach was similar, although less complex in that the values of CVR_{LL} and CVR_{UL} were chosen to be 50% and 145% of CVR_{SP} respectively. This meant that the $MABP$ could be directly set at 100% and the autoregulation mechanism given a period of 20s to allow CVR to alter in the porous blocks in order to reach CBF_{SP} in every efferent artery. Once CBF_{SP} had been achieved, the upper and lower limits and the set point for CVR were known.

It should be noted that the actual numerical values of CVR obtained for the set point and lower limit of autoregulation bear no resemblance to the concept of CVR which would hold in the relation $CPP = CBF \cdot CVR$ and should not be interpreted as such. In fact it should be apparent that dimensionally they are not even equivalent. The values of CVR incorporated into the body force term of (7.3) would be best likened to the reciprocal of the permeability in the Darcy's law [33] where their actual effect of the CBF is also is dependent upon the lengths and diameters of the porous blocks. The use of CVR in this way presents an elegant way to couple the 1D system of ODE's that the autoregulation model produces with the 3D form of the governing equations.

All of the circle of Willis simulations were performed using the commercial finite volume solver *Fluent*, using the segregated-implicit solver, with node based gradient reconstruction, second order temporal discretization, second order upwinding, SIM-PLE pressure velocity coupling, with a convergence tolerance of 10^{-5}. The AMG method used the V cycle for the pressure correction equation and the flexible cycle for the momentum equations, using the default parameters for the number of pre and post relaxation sweeps and the termination criteria (see Appendix C for an explanation of these solution parameters). When the non pulsatile simulations were run, the timestep size was set at 0.2s with 100 iterations per timestep. When the pulsatile simulations were performed the timestep size was set at 0.02s, such that fifty points per cardiac cycle would be specified, with 50 iterations per timestep. The idealized circle of Willis model simulations were performed on a single desktop PC, running *serial* Fluent on a *Windows XP* operating system, each requiring approximately 2 days to complete. The patient specific circle of Willis models were performed on an *IBM power5-575* HPC running an *AIX* 5.3 operating system using parallel Fluent on sixteen processors, with each requiring approximately 10 hours to complete. The autoregulation model, pressure boundary conditions and various other functions relating to their functioning were incorporated into Fluent as a *compiled user defined function*, which creates a *shared object library* to be called at runtime during the solution procedure.

Chapter 8

Idealized Circle of Willis Results

8.1 Introduction

The reason for the development of the idealized circle of Willis models was to attempt to draw some conclusions about how certain anatomical configurations and variations in arterial diameter for a given configuration affect the flow patterns under conditions of both normal and reduced afferent perfusion pressure. Three different anatomical configurations were modelled (as illustrated in Chapter 5), namely a complete Circle of Willis, a Fetal P1, and a Missing A1, as these comprised some of the more common pathological conditions [13]. Within each configuration the effects of a reduction in diameter of a single arterial segment was simulated and the subsequent collateral flow patterns around the CoW examined. Using the measurements in Table 5.1, each arterial segment was reduced by one standard deviation of its mean diameter. To limit the number of numerical simulations to a manageable task, only three arterial segments per configuration were reduced and were chosen based on their assumed importance in providing collateral flow (given the anatomical configuration) to provide the most detrimental scenarios in terms of maintaining cerebral perfusion. In order to test the ability of the CoW to reroute blood flow a unilateral 20mmHg pressure drop was imposed in an ICA, similar to that employed in the thigh cuff studies of Newell et al [112]. Similar to the arterial choice, the selection of which ICA to impose the pressure drop in was dependent on the configuration and chosen to provide the most detrimental effect on providing collateral flow. For lack of better information it was decided to utilize the same set point flowrates for all three configurations, thereby assuming that the presence of an anatomical variation does not affect how much blood is supplied through an efferent artery. Since the autoregulation model employed was designed such that CVR did not react over the cardiac cycle, it was also decided to ignore the pulsatile effects of blood and observe the changes in mean CBF throughout the CoW.

At the time of these simulations, computer resources were limited so the computational meshes were as fine as possible, while still allowing the grid and flow field data to be stored on a single desktop PC with 1Gb of memory. The computational meshes for the models contained approximately 1 million tetrahedral volumes, but mesh convergence was not investigated for these models.

Before presenting the results the sign convention for the direction of flowrates must be set. While for the afferent and efferent arteries, the positive flowrates are into and out of the CoW respectively, and don't reverse direction, the flow through the communicating arteries does reverse direction. The convention used for these results is that flow through the PCoA's is defined as positive in the posterior direction (i.e. blood that is supplied by the ICA and travels to the PCA through a PCoA is positive) and blood flow through the ACoA is defined positive when moving from the left to the right side of the CoW.

8.2   Complete Circle of Willis

For the complete circle of Willis the arteries chosen for reduction were the three communicating arteries (LPCoA, RPCoA and the ACoA) as it was thought that these would have the greatest affect on providing collateral circulation. As the complete CoW model was relatively symmetrical, the choice of ICA in which to impose the pressure drop was arbitrary and so the RICA was chosen. Figures 8.1(a),8.1(b) and 8.1(c) illustrate the transient results through the efferent, communicating and afferent arteries of the normal complete circle of Willis respectively. All of the simulations showed a similar transient response and so for this configuration and the other two, only the transient results for the normal case of each configuration will be presented. It can be observed in Figure 8.1(c) that under normal conditions an equal amount of blood flow is supplied through the ICA's (approximately 270mL/min) which is significantly more than that supplied from the posterior of the CoW (approximately 220mL/min through the BA). Concerning the communicating artery flow (Figure 8.1(b)), it can be observed that under normal conditions there is essentially no flow through the ACoA and a small equal amount (approximately 20mL/min) through each PCoA from anterior to posterior.

Following the pressure drop it can be observed that the effects are transmitted rapidly through the CoW and there is a reduction in efferent blood flow through all of the efferent arteries, although the ipsilateral RMCA is most affected as it is supplied entirely by the RICA (Figure 8.1(a)). The rapid pressure drop also causes a rapid increase in the flow through the communicating arteries (except for the contralateral LPCoA) which does not alter for the remainder of the simulation. It can also be observed that during the next 20 seconds after the pressure drop, the autoregulation mechanism is activated and causes a reduction in CVR in all of the efferent arteries to restore CBF to its set point value. Furthermore, since the flow through the communicating arteries does not increase following its initial 'jump', the increase in supply is made up of continuing increases through the afferent arteries, but mainly the RICA. It can be observed that the dynamics closely follow those observed by Newell et al, and in keeping with the autoregulation model it can also be noted that because the perfusion pressure drops throughout the CoW, the set point flowrate also decreases, meaning that although the flowrates are restored to their set point value, they are not equivalent to before the pressure drop (most noticeable between the two MCA's). The streamline plot of Figure 8.1(d) illustrates the high velocities through the ACoA, the ipsilateral PCoA, and P1 segment of the ipsilateral PCA as the blood flow is rerouted through the CoW. One of the interesting results is that the amount of flow through the A1 segment of the ipsilateral ACA is reduced essentially to zero. Instead the supply to the RACA is provided through the ACoA from the left side of the CoW, since all of the remaining supply through the RICA is used to perfuse the RMCA.

The reductions in arterial diameters for the complete CoW did not cause an appreciable difference in flow patterns. Because the flowrates through the efferent arteries are essentially the same for every variation, they will not be presented, though the variations in afferent and efferent flowrates will. Figures 8.2(a) and 8.2(b) illustrate the effect of the variations in diameter with the steady state flowrates both before and after the pressure drop for the afferent and communicating arteries respectively. It can be seen that the RICA pressure drop causes an average 60% decrease through the RICA (although if the autoregulation mechanism were not present it would be greater) and approximately a 25% increase through the LICA. Furthermore there is an equal increase in each VA of approximately 50%. While the flow through the contralateral LPCoA is negligibly affected it can be observed

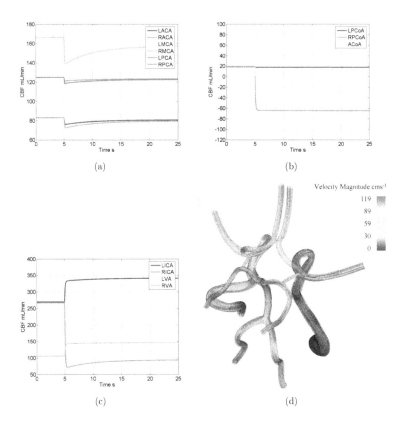

Figure 8.1: Transient flowrates through the (a) efferent arteries, (b) communicating arteries, (c) afferent arteries, and (d) a streamline plot through the complete circle of Willis at the end of the simulation.

that the flow through the ipsilateral RPCoA reverses direction and essentially triples in magnitude (apart from the case where it is reduced in diameter, where it only slightly more than doubles) as flow is rerouted from the posterior supply to the anterior supply. It can also be observed that the large increase in flow through the ACoA is not affected significantly in the case where it is reduced in diameter. The most significant variation is the reduction in the ipsilateral RPCoA which decreases the amount of flow which may be rerouted from the posterior supply, resulting in a reduced collateral flow that is compensated via and increase in supply through the RICA.

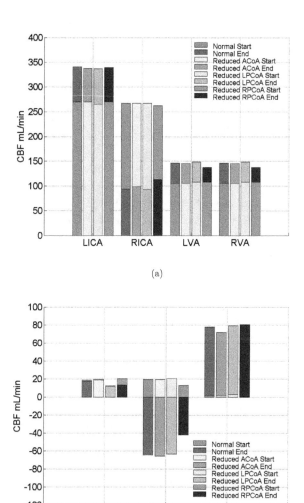

(a)

(b)

Figure 8.2: Steady state (a) afferent and (b) communicating flowrates through the complete circle of Willis with its arterial variations, both under normal cerebral perfusion pressure conditions and after the pressure drop. Note: the lighter shade of a bar illustrates the flowrate under normal conditions (denoted 'Start') and the darker shade of a bar illustrates the flowrate in response to the pressure drop at the end of the simulation (denoted 'End').

8.3   Fetal P1 Circle of Willis

For the fetal P1 circle of Willis the arteries chosen for reduction were the ACoA, the A1 segment of the ipsilateral ACA (the RACA in this case) and the P1 segment of the RPCA (the fetal P1). In this case the ICA chosen to have the pressure drop imposed was the ipsilateral RICA, as this would require flow to be rerouted through the fetal P1 and around the anterior of the CoW. Figures 8.3(a), 8.3(b) and 8.3(c) illustrate the transient flowrates through the efferent, communicating and afferent arteries respectively for the normal case of the fetal P1. It can be observed that the dynamics of the response to the pressure drop are extremely similar to the complete CoW. One interesting result is that under normal conditions there is significantly more flow through the ICA's compared to the complete CoW (approximately 300mL/min compared to 270ml/min) and slightly more through the ipsilateral RICA compared to the LICA. There is also significantly less supply through the posterior compared to the complete CoW (approximately 160mL/min compared to 220mL/min). The reason for the large reduction in posterior supply is that since the RPCA can essentially be thought of as originating at the ICA instead of the BA (and the fetal P1 is the communicating artery), much more of its efferent supply comes from the ICA (hence its aforementioned increase in supply) and therefore less is required from the BA. In addition the reduced diameter of the fetal P1 makes it more difficult for afferent supply from the BA to perfuse the RPCA. Therefore a much larger flow exists through the ipsilateral RPCoA, but only slightly more flow through the contralateral LPCoA, both of which are in the positive anterior to posterior direction.

The streamline plot of Figure 8.3(d) illustrates the higher velocities through parts of the CoW in response to the pressure drop compared to the complete CoW, the highest being through the fetal P1. Following the pressure drop in the RICA (Figure 8.3(c)), despite the gradual increase over the course of the simulation, not enough afferent supply can be drawn through it to restore supply to the RMCA, meaning that the only other source is via the fetal P1. This is also exhibited by the fact that the flow through the ipsilateral RPCoA reverses direction, meaning that blood flow from the BA is being used to feed the RMCA as well as make up the whole supply of the RPCA. While the efferent supply to the RPCA and RMCA can be restored, it requires quite high velocities through the fetal P1. It can also be observed that as with the complete CoW, following the pressure drop there is essentially no flow through the A1 segment of the RACA.

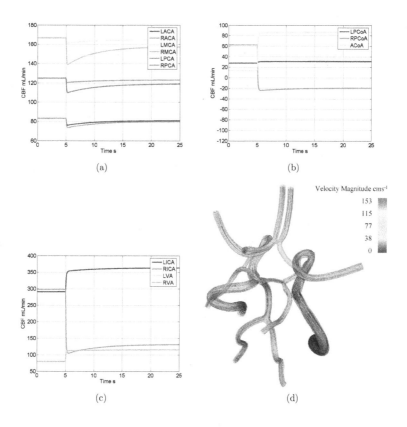

Figure 8.3: Transient flowrates through the (a) efferent arteries, (b) communicating arteries, (c) afferent arteries, and (d) a streamline plot through the fetal P1 circle of Willis at the end of the simulation.

The reductions in arterial diameters exhibited are more significant for this configuration compared to the complete CoW. Figures 8.4(a) and 8.4(b) illustrate the steady state afferent and communicating flowrates respectively. It can be observed that there is approximately a 20% increase through the contralateral LICA, which is slightly less than for the complete CoW. In addition, the reductions through the ipsilateral RICA are only approximately 50% and the increase through the VA's approximately 37% all of which are also less than that for the complete CoW.

For a normal fetal P1, the flow in the A1 segment of the RACA is directed toward its A2 segment, but after the RICA drop it reverses direction (as flow is rerouted through the ACoA) to contribute supply to the RACA and RMCA. For all other variations however, the flow is in the reverse direction under normal conditions, meaning that blood is rerouted from the LICA through the ACoA to perfuse the RMCA and supply the RPCA. After the pressure drop the flow reverses direction such that blood supply from the RICA along the A1 segment of the RACA travels towards the A2 segment. One of the more significant variations is the reduced diameter of the fetal P1 itself. The most noticeable effect is that for all other cases, the pressure drop causes the flow through the RPCoA to reverse direction as it is normally used to supply the RMCA and the flow through the fetal P1 segment makes up supply to the RPCA. In this particular case however the flow is higher under normal conditions, but does not reverse direction following the pressure drop. This is because not enough flow can be rerouted through the fetal P1 segment to completely supply the RPCA and so to make up supply to the RMCA there is a very large increase in flow through the ACoA which contributes supply to the RACA and RMCA.

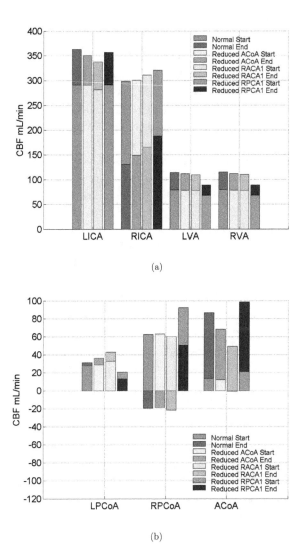

Figure 8.4: Steady state (a) afferent and (b) communicating flowrates through the fetal P1 circle of Willis with its arterial variations, both under normal cerebral perfusion pressure conditions and after the pressure drop. Note: the lighter shade of a bar illustrates the flowrate under normal conditions (denoted 'Start') and the darker shade of a bar illustrates the flowrate in response to the pressure drop at the end of the simulation (denoted 'End').

8.4   Missing A1

For the missing A1 circle of Willis the arteries chosen for reduction were the A1 segment of the LACA and both of the PCoA's. For this configuration the ICA chosen for the pressure drop was the contralateral LICA, as it is the major source of blood supply to both ACA's in addition to the LMCA and therefore reductions in its supply would require blood flow to be rerouted from the posterior and right anterior supply. It should be noted that this is the only one of the three idealized models where the circle of Willis is completely broken. Figures 8.5(a), 8.5(b) and 8.5(c), illustrate the transient response for the normal case. As with the previous two cases the dynamics are similar, the major difference being that under normal conditions there is significant difference in the afferent supply through the ICA's with approximately 100mL/min more through the LICA compared to the RICA. This is of course expected since the supply to the RACA is coming from the LICA instead of through the RICA. Flow through the VA's is similar to the other two cases, but the RPCoA flowrate is generally much higher under normal conditions (approximately twice as much) but is still in the positive direction since blood supply from the RICA is used to perfuse the RPCA. The flow through the LPCoA is negative under normal conditions, meaning that the afferent supply from the posterior contributes to the supply of the left anterior part of the CoW.

Following the pressure drop it can be observed that there is a significantly large increase in the flow rerouted through the contralateral LPCoA (the largest flow through the LPCoA for all of the configurations) from posterior to anterior. There is also an increase in the flow through the RPCoA, but nowhere near as significant. The streamline plot of Figure 8.5(d) illustrates the high velocities through the P1 segment of the LPCA as this flow is rerouted following the pressure drop. While the maximum velocity is not as high as that in the fetal P1 case, it is higher than that through the complete CoW.

Figures 8.6(a) and 8.6(b) illustrate the steady state flowrates through the afferent and communicating arteries respectively for all variations. It can be observed that for all cases there is the smallest decrease in afferent supply through the artery with the imposed pressure drop, averaging approximately 40% compared to 60% and 50% for the complete CoW and fetal P1 respectively. As the circle is broken with this configuration, there is only one path for collateral flow and despite the

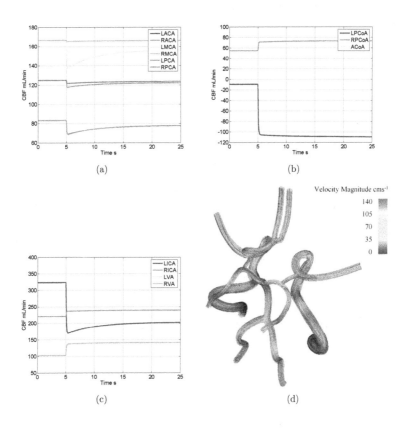

Figure 8.5: Transient flowrates through the (a) efferent arteries, (b) communicating arteries, (c) afferent arteries, and (d) a streamline plot through the missing A1 circle of Willis at the end of the simulation.

fact that the PCoA's of the missing A1 cases tend to be larger than those with a complete CoW, it does not make up for the fact that the circle is disconnected. Consequently, to restore efferent blood supply to the LMCA and the two ACA's, much of the afferent blood supply has to be drawn through the LICA. There is also a smaller increase in afferent supply through the contralateral ICA of approximately 9% compared to the 25% and 20% for the complete CoW and fetal P1 respectively, which is due to the fact that the resistance associated with rerouting blood flow from the right to the left side of the CoW mean that blood supply from the RICA will not be directed toward the LMCA and ACA's. Instead it is used only to maintain

supply to the RPCA, while more of the afferent flow from the BA is rerouted to the
LMCA and ACA's.

It can be observed that the reduction of the A1 segment of the LACA has no
impact on the flow patterns throughout the CoW. It is not part of a collateral
pathway, and in order to restore efferent supply to the ACA's the same amount of
supply must be rerouted through it (and hence through the LICA and LPCoA). One
of the more interesting and significant variations is the reduction of the contralateral
RPCoA. As can be observed in Figure 8.6(b) the reduction of the RPCoA, while
not significantly altering the flow through itself both under normal conditions and in
response to the pressure drop, causes a large reduction in flow through the LPCoA
supplying the LMCA and ACA's. The reduced diameter means that less afferent
supply from the RICA can be used to supply the RPCA and consequently this has
to be compensated via the afferent supply from the BA's. The interesting point is
that following the pressure drop there in not a large increase in afferent flow through
the BA to meet both the needs of the RPCA and reroute blood flow through the
LPCoA. Instead the amount of flow rerouted through the LPCoA is much less and
the restoration of efferent supply to the LMCA and ACA's is achieved through
drawing more afferent supply through the LICA.

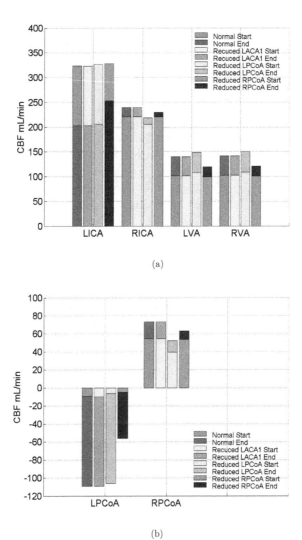

Figure 8.6: Steady state (a) afferent and (b) communicating flowrates through the missing A1 circle of Willis with its arterial variations, both under normal cerebral perfusion pressure conditions and after the pressure drop. Note: the lighter shade of a bar illustrates the flowrate under normal conditions (denoted 'Start') and the darker shade of a bar illustrates the flowrate in response to the pressure drop at the end of the simulation (denoted 'End').

The results from the twelve simulations presented illustrate that different CoW configurations and variations in diameters of the various arterial segments comprising the CoW do affect the amount of blood flow through afferent and communicating arteries under normal conditions, and influence the manner in which collateral flow will be provided in response to a reduction in perfusion pressure. Furthermore, the effects of the variations in arterial diameter on collateral flow patterns are more significant when there is an anatomical variation present in the circle. The effects of these variations were probably not as pronounced as they could have been, since the 20mmHg pressure drop was not a large enough reduction in perfusion pressure that the autoregulation mechanism couldn't restore afferent blood flow. If a larger pressure reduction or complete occlusion of one or more of the afferent arteries were imposed then the more detrimental variations would become apparent. From the anatomical measurements of Table 5.1 it appears that when an anatomical variation is present, the communicating arteries tend to be larger, providing a greater capacity for collateral flow, but whether or not this can counteract the effects of the anatomical variation are not shown by the simulation results. When the capacity for collateral flow is exhausted and the only way that the efferent arteries may restore their set point flowrates is to draw more afferent blood supply from the reduced perfusion pressure artery, then this illustrates a case where further reductions will lead to a scenario in which blood flow cannot be restored to all cerebral territories. The increase in afferent supply comes from the reductions in peripheral resistance of the efferent cerebral territories, but as the perfusion pressure is further reduced, the CVR on these efferent territories will reach its lower limit (maximum vasodilation) and will not be able to draw any more afferent blood supply. This effect can be observed in the present results for the missing A1, which has the smallest steady state reduction in afferent supply through the pressure drop imposed ICA, as it is the only option to restore efferent CBF. It would therefore be expected that given further perfusion pressure reductions, this configuration would be the first to reach its lower autoregulatory limit and result in the incomplete restoration of efferent CBF.

The results highlight the difficulty of making generalized conclusions about how configurations, variations in arterial diameter and losses in perfusion pressure in the afferent arteries affect the overall flow patterns, for the simple reason that there exist so many variables. The simulations presented here only included the variation of one arterial segment at a time, with a reduction of only one standard deviation. Furthermore the imposed pressure drop was only in one artery. It is not difficult to see that including variations of multiple arterial segments by one or more standard deviations at a time, with pressure drops in one or more afferent arteries of varying magnitude, would result in an insurmountable number of simulations being required. Even then processing the data in a manner which could be used to make generalized conclusions would be a very difficult task. There is also the result illustrated in Table 6.2, that normal flowrates through the efferent and afferent arteries of the CoW (and most likely communicating flowrates, though they were not measured in the study) do vary between individuals, which further complicates the problem in that the significance of the amount of collateral flow that can be provided by an individual CoW is relative to the normal efferent cerebral flowrates for an individual.

It is for these reasons that further simulations and enhancement of the idealized models has been abandoned and simulations on patient specific models adopted. As will be addressed in the next chapter, the question then becomes not, how much does a variation in this part of the CoW affect the flow patterns, but rather, how effective is a given patient's CoW for providing collateral flow in response to a loss in afferent perfusion pressure? While comparisons can still be made between anatomical variations, this will not be done by controlled variations in arterial diameters (since these are defined implicitly in the MRI data), but by a number of simulations on different patient specific geometries. The major application of the understanding that comes from idealized models would be answering clinical questions related to patient specific cerebral perfusion problems, so it naturally follows to develop the technique to create patient specific models and thereby answer these patient specific questions and collect data from a number of such simulations to make generalized conclusions, rather than continue to pursue work with idealized models.

Chapter 9

Patient Specific Circle of Willis Results

9.1   Introduction

Apart from the development of using geometry obtained directly from MRI data, the patient specific circle of Willis models incorporate a number of extensions upon the idealized models in terms of the CFD simulations, such as the adaptive cartesian grid generation technique and the variation of afferent arterial blood pressures over the cardiac cycle. Since the available computing power for these models was not a limiting factor (as it was for the idealized models) some investigation into mesh convergence was performed. Before presenting the results of combinations of occlusion and stenosis for the three patient specific models investigated in the present study, the results of the mesh convergence simulations, a comparison between pulsatile and non-pulsatile simulations, and a comparison between the assumption of a Newtonian and non-Newtonian viscosity will be presented.

9.1.1   Mesh Convergence

Both the idealized and patient specific model simulations are designed to adjust the body force term in the porous blocks of the efferent arteries such that the correct value of CBF, and hence C_tCO_2 for the patient specific models is achieved, and this is the case regardless of the level of refinement of the computational grid. It is important to note however that the flow field is being adequately resolved, independent of the presence of the autoregulation mechanism. The major results of the simulations are the efferent flowrates (and variables such as OEF, C_tCO_2 and O_2 delivery which depend upon the efferent flowrates) following the incorporation of a stenosis or occlusion of an afferent artery, which depends partly on the increase in afferent blood supply, but also on the ability of the communicating arteries to provide collateral flow.

It was decided that the most appropriate way to judge whether or not the mesh had been refined to an appropriate level was to perform a number of simulations on the same patient specific circle of willis geometry, with varying levels of mesh refinement and imposing non-pulsatile afferent and efferent pressures with the autoregulation model 'turned off'. Furthermore the afferent and efferent pressures would be asymmetrical in order to cause flow through the communicating arteries. Mesh convergence would then be based on two criteria; the first being percentage difference in communicating artery flowrates between a given level of mesh refinement and the previous coarser mesh, and the second being the velocity profiles at the beginning of the LMCA immediately after the ICA - ACA junction. The arbitrary choice made for the initial mesh was a total of approximately fifty thousand cells and this number as essentially doubled between every subsequent simulation.

Figure 9.1(a) illustrates the results of the mesh convergence simulations in terms of the communicating artery flowrates. The circle of Willis model selected for the simulations was the complete circle of Willis and there were hence three communicating arteries within which the flowrates were monitored and three percentage difference calculations to compare between simulations. The percentage difference presented in Figure 9.1(a) between sequential levels of mesh refinement are the maximum percentage difference of the three arteries. It can be observed that there are large differences in communicating artery flowrates for meshes comprising a total number of cells less than one million, although subsequent levels of refinement exhibit increasingly smaller effects on the amount of collateral flow. The maximum percentage difference between meshes comprising of one million and two million cells is approximately 0.5% and so based on the first criteria a mesh of approximately two million cells would yield an accurate solution.

Figure 9.1(b) illustrates the results of the mesh convergence simulation in terms of the proximal LMCA velocity profiles. It can be observed that refining the mesh increases the peak velocity through the artery and similar to the communicating artery flowrates it can be observed that with meshes comprising of less than one million cells there is a difference of approximately 5cm/s in peak velocity. With the level of mesh refinement numbering between one and two million cells it can be seen

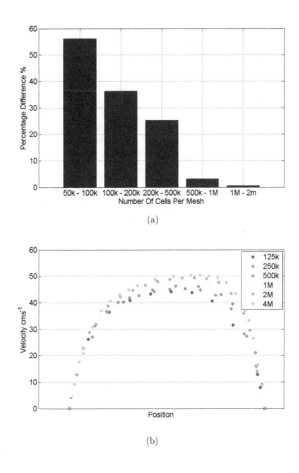

Figure 9.1: (a) The maximum percentage difference in communicating artery flowrate between successive mesh refinements of the complete circle of Willis model, (b) the LMCA velocity profile immediately distal to the ICA-ACA junction for various levels of mesh refinement.

that peak velocity differs by approximately 2cm/s and between two and four million cells it differs by approximately 1cm/s. Based on the constant cross sectional area of the vessel a difference in peak velocity of 1cm/s would result in a percentage increase in flowrate of approximately 2% which would be a reasonable error to allow and assert that the solution had reached mesh independence. It should be noted that since the flowrates in the stenosis and occlusion simulations are dependent upon the body force term in the porous block, the effects of utilizing an overtly coarse mesh would be to decrease the maximum flowrate which could be restored in an efferent artery upon reaching its lower limit of autoregulation. Since the magnitude of the error associated with a mesh size of approximately two million cells appears to be of the order of a few percent, meshes of this size are considered to give a sufficiently accurate solution.

9.1.2   Pulsatile Effects

To investigate the significance of the pressure pulse on flow patterns throughout the circle of Willis, and furthermore how the pulse affects the autoregulation mechanism, two test simulations were performed on the complete circle of Willis model; one pulsatile and the other non-pulsatile. In both cases the three afferent arteries were subjected to a reduction in afferent blood pressure of 20mmHg (Figure 9.2(a)) such that CBF would be reduced in all of the efferent arteries and the autoregulation mechanism would respond by reducing CVR. To illustrate an example result, Figures 9.2(b) - 9.2(f) show the transient responses of the CBF, C_tCO_2, CVR, OEF and O_2 delivery in the LACA for both the pulsatile and non-pulsatile simulations, although results were similar in the other efferent arteries.

The most important result of these simulations is that although CBF varies over the cardiac cycle (as well as OEF and O_2 delivery) for the pulsatile simulations, they follow the cardiac cycle averaged values of the non-pulsatile simulation. Another important result is that in response to the pressure drop, the autoregulatory response in terms of a reduction in CVR is essentially the same (Figure 9.2(d)). This is as would be expected since the gains for the autoregulation activation were set to match the data of Newell et al [112] which would mean that CVR would not be able to alter appreciably over the cardiac cycle, but only to changes lasting over many cardiac cycles. It can be observed in Figure 9.2(e) that OEF varies over the cardiac cycle and the interpretation of this result requires some discussion. The gain for the oxygen extraction activation was chosen such that OEF could alter more rapidly than CVR as it was postulated that physiologically in the cerebral tissue there should be little delay in a reduction in CBF and an increase in the oxygen being extracted from the blood to supply the cerebral tissue. Since CBF varies over the cardiac cycle then so too does the O_2 delivery, and as a result, less oxygen has to be extracted from the arterial blood to meet the metabolic demands of the

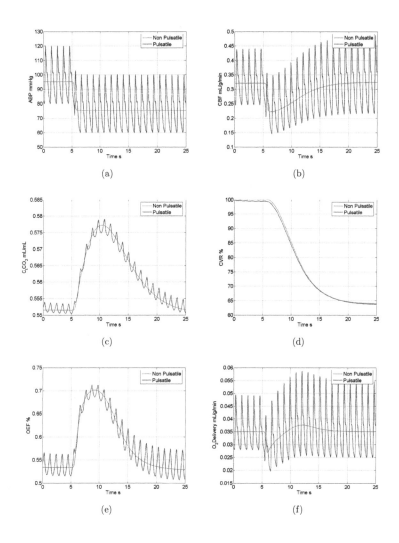

Figure 9.2: An investigation into the pulsatile effects following a 20mmHg drop in afferent pressure for the complete circle of Willis model, illustrating for the LACA (a) the afferent pressure in the LICA, (b) the transient CBF, (c) the transient $C_t CO_2$, (d) the transient CVR, (e) the transient OEF and (f) the transient O_2 delivery.

cerebral tissue, hence the variation in OEF over the cardiac cycle. Physiologically, this situation is unlikely to occur because at the level of the cerebral capillaries, where the exchange of oxygen and carbon dioxide is occurring, the pulsatile nature of the blood flow has been dampened into what is essentially steady flow. In order to allow for OEF to alter in cases of reduced CBF and on a timescale shorter than that of the CVR, it is necessary with the present model to accept the alteration of OEF over the cardiac cycle. The same result occurs for the tissue carbon dioxide concentration varying over the cardiac cycle (Figure 9.2(c)) and there is the same physiological argument that at the capillary level where the transport of carbon dioxide into the bloodstream is occurring, the pulsatile nature of the flow would be damped out. Again this result remains a limitation of the way in which the autoregulation model has been implemented, but the important result is that the cardiac averaged changes in these parameters do show the correct responses.

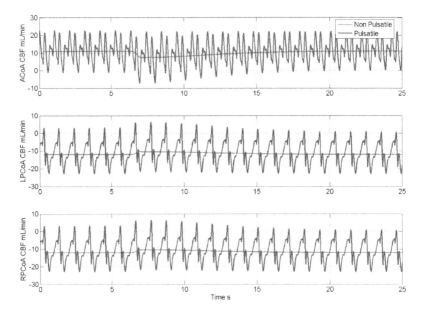

Figure 9.3: Transient flowrates through the three communicating arteries of the complete circle of Willis model, illustrating a comparison between the pulsatile and non-pulsatile flowrates.

The other important result of these simulations is the effect of the pulsatile nature on the blood flow on the communicating artery flowrates. Figure 9.3 illustrates the pulsatile and non-pulsatile flowrates through the three communicating arteries of the complete Circle of Willis. It can be observed that while the efferent flowrates

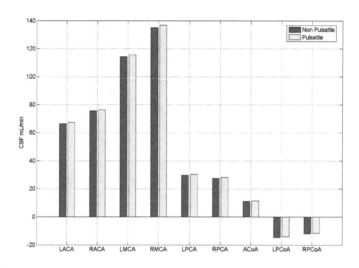

Figure 9.4: A comparison between the non pulsatile and cardiac cycle integrated pulsatile flowrates in the efferent and communicating arteries, at the end of the 20mmHg afferent pressure drop simulation.

of the pulsatile simulation (corresponding to the LACA example in Figure 9.2(b)) follow a profile similar to the pressure waveform in Figure 9.2(a), the waveform of the communicating artery flowrates is much more complex. The reason for this complex flowrate waveform is most likely due to the slight differences in the phase and the shape of the pressure profiles on each of the afferent arteries, which would result in a complex variation in the pressures occurring at the two ends of a communicating artery. It can also be observed that all three of the communicating artery flowrates exhibit blood flow which reverses direction over the cardiac cycle. Since the drop in ABP was the same in all afferent arteries, the communicating arteries were not used significantly to provide collateral flow and this is evident via the minimal changes in flowrate.

Figure 9.4 illustrates a comparison between the efferent and communicating artery flowrates at the end of both simulations, where for the pulsatile simulations the end values were time averaged over the last cardiac cycle of the simulation. The major result is that as can be observed in the plot, there is very little discrepancy between the efferent and communicating artery flowrates with a maximum difference of approximately 5% occurring in the LPCoA. In terms of the oxygen delivery to the cerebral tissue the maximum difference calculated based on this simulation was 3%. Since it is only the cardiac averaged results of CBF and O_2 delivery which are desired to understand the clinical impacts of stenosis or occlusion in a given circle of Willis geometry, the implication of these simulations is that the stenosis and occlusion simulations can henceforth be treated as non-pulsatile, reducing the simulation time by a factor of ten, but still retaining the accurate measures of these cardiac averaged autoregulatory variables. Since the Womersley number associated with these simulations is approximately 3, it can safely be assumed that the non-pulsatile reduction in perfusion pressure through a stenosis, would be approximated by the cardiac averaged loss in perfusion pressure of a pulsatile simulation.

9.1.3   Non Newtonian Effects

Since it is generally assumed that the behavior of blood is essentially Newtonian in the larger arteries of the cardiovascular system, it was decided to investigate the effects of the incorporation of the Carreau-Yasuda viscosity model. The reason for interest lies in the fact that due to the smaller diameters of the communicating arteries, the amount of blood flow through them is generally much smaller than compared to the other cerebral arteries, which would correspond to lower shear rates and hence a higher apparent viscosity of the blood within them. There is the potential for the non-Newtonian properties of blood to potentially affect the amount of collateral flow through a communicating artery. The non-pulsatile simulation described previously (which had a Carreau-Yasuda model incorporated) was repeated with a Newtonian viscosity incorporated, at the infinite shear viscosity of 0.00348Pa.s. Figure 9.5 illustrates the comparison between simulation in terms of the amount of collateral flow and it can be seen that in fact the non Newtonian properties of blood have no effect on the amount of blood flow provided through the communicating arteries.

As a further illustration, Figure 9.6 illustrates the contours of viscosity over the boundary of the complete circle of Willis. It can be observed that throughout the majority of the model, including throughout the communicating arteries, the viscosity is at its infinite shear value. The only exceptions to this occur in the arteries around the circle which were truncated and 'capped' so that no flow would pass through them. In this case stagnant flow results in the viscosity being at the

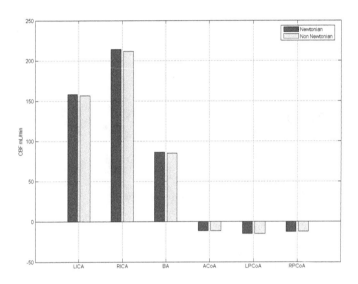

Figure 9.5: A comparison between the Newtonian and non Newtonian viscosity models on communicating artery flowrates following a 20mmHg afferent pressure drop for the complete circle of Willis model.

zero shear value, but this has essentially no effect on the flow patterns throughout the CoW. Furthermore it can be seen that there are very small local variations in viscosity, particularly around the walls of the ACA's. These variations are a result of the 'lumpy' surface topology, which is a general feature resulting from the incorporation of patient specific geometry, but having very little effect on the overall flow patterns.

The incorporation of occlusion and stenosis in the afferent arteries of the three patient specific models results in a large number of possible simulations which could be performed, when the varying degrees of stenosis and their possible combinations are considered. It was decided that in order to limit the simulations to a manageable number only 50% stenosis, 70% stenosis, and 100% stenosis (i.e. complete occlusion) would be considered.

Initially, simulations were performed with a unilateral stenosis of 50% and then

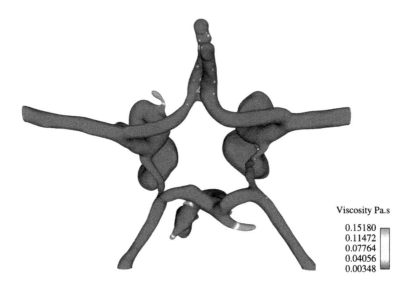

Figure 9.6: A contour plot of the Carreau-Yasuda non Newtonian viscosity throughout the complete circle of Willis model.

70% imposed on a single afferent artery for each patient specific geometry, but the results showed that even with the higher 70% stenosis, its effect on perfusion pressure through the CoW and hence on the efferent flowrates was negligible, and this was the case for all patient specific models. In each case efferent CBF could be completely restored with only a very minimal decrease in CVR for the efferent arteries ipsilateral to the stenosis, meaning that C_tCO_2, OEF and O_2 delivery were all at their normal levels at the end of the simulations. For this reason the results of a single stenosis will not be presented and instead the combinations of a unilateral occlusion with a contralateral stenosis will, as they present some more interesting results.

For each patient specific geometry investigated, three groups of simulation results are presented. Within each group a single afferent artery is occluded and one of the remaining two afferent arteries has an imposed stenosis of 50% or 70%. With the patient specific autoregulation model there is an increase in the number of autoregulatory variables that is associated with a given efferent artery. Since the primary concern are their end values after CVR and OEF have responded to the stenosis to maintain oxygen delivery, it was decided that the best way to present the simulation results is via the radial plots (Figures 9.7(a) - 9.7(e)). In these plots the vertex of each hexagon displayed on the radial plot represents the steady state value of a particular autoregulatory variable in the corresponding efferent artery. The hexagons labelled 'baseline' represent the values of the autoregulatory related properties in the unstenosed, normal case and the dashed hexagons for the CVR and OEF plots illustrate the lower and upper limits for these two variables respectively.

9.2 Complete Circle of Willis

The results of the PC MRI scans in terms of the cardiac averaged efferent artery flowrates for the complete circle of Willis are presented in Table 9.1 along with the corresponding values of CVR when the model was in its set point state. The total efferent blood flow in this case is 448.9mL/min which was the lowest found among the three volunteers and significantly lower than the generally accepted average value of 750mL/min. It is important to note however that the generally accepted value accounts for total cerebral blood flow, whereas the measurements made for this study only include the major efferent arteries, neglecting the flowrate through a number of other arteries in the brain proximal the circle of Wills. It would therefore be expected that the total measured flowrate be somewhat different to the standard value, but in any case these measurements are within an acceptable physiological range [127, 165]. As illustrated in Figures 9.7(c) and 9.7(d) the normal levels of carbon dioxide and oxygen extraction in the tissue are approximately 55mL/100mL and 54% respectively.

Table 9.1: The efferent flowrates for the complete circle of Willis measured with PC MRI, and the values of CVR required to achieve them in the numerical simulation.

	LACA	RACA	LMCA	RMCA	LPCA	RPCA
CBF mL/min	66.5	75.6	114.1	135.1	29.9	27.7
CVR mm^{-2}	1673	1967	998	862	4312	3933

9.2.1   LICA Occlusion

The results for the autoregulatory response to the LICA occlusion in combination with a unilateral stenosis are presented in Figures 9.7(a) - 9.7(e). It can be observed in Figure 9.7(a) that the effects of the LICA occlusion and the other stenoses result in efferent CBF not being completely restored in the LACA and LMCA ipsilateral to the occlusion as CVR in these arteries approaches the lower limit of autoregulation (Figure 9.7(b)). As a result, C_tCO_2 for these two arteries remains elevated following the simulations. An important point to note is that CVR did not in fact reach the lower limit because the 20s allowed for the simulation was not enough time for CVR to reduce to the lower limit. This result was found in a number of simulations and occurs because the CVR dynamic is essentially a proportional controller and as the CBF returns to its normal level, then so diminishes the drive for a further reduction in CVR. Given more simulation time, these two efferent flowrates may in fact return to their normal value, but the important result is that as can be observed in Figure 9.7(d), there is only a small increase in OEF required to maintain O_2 delivery to these arteries, which is the major result of the simulation. In cases where the occlusion and stenosis result in a very dramatic reduction in CBF, the lower limit of autoregulation will actually be reached and this will occur very quickly. It can also be observed that the 70% RICA stenosis has a much greater effect on the efferent flood flow through the contralateral RACA and RMCA than the 50% stenosis, illustrated by the much larger reduction in CVR required to restore the efferent CBF to their normal levels. This is a finding that occurs in a number of simulations, indicating that quite a large stenosis is required (even when in combination with an occlusion), in order to affect cerebral blood flow appreciably.

The afferent and communicating artery flowrates (Figure 9.8) show that under normal conditions the communicating artery flowrates are all approximately 15mL/min. In keeping with the sign convention for the direction of communicating artery flowrate from the previous chapter (i.e. positive ACoA flowrate is from left to right and positive PCoA flowrate is from anterior to posterior), it can be seen that the two PCoA flowrates differ compared to the idealized simulation of a complete circle of Willis in that here the direction of flow is from posterior to anterior. Furthermore, because of the asymmetry of the complete CoW in terms of geometry and normal ACA flowrates it can be observed that there is approximately 10mL/min of blood flow from right to left, which differs from the idealized complete CoW where the symmetry of the geometry and ACA flowrates resulted in essentially no flow

through the ACoA. In response to the occlusion and stenoses it can be observed
that the approximately 150mL/min normally supplied by the LICA is generally restored by an almost equal increase from the RICA and the BA, the exception being
in the case of a 70% BA stenosis where the majority of the afferent supply to the
LACA and LMCA originates from the RICA. In keeping with this result there is
a relatively equal increase in the amount of collateral flow provided by the ACoA
and the ipsilateral LPCoA, between approximately 65 and 100mL/min, but similar
to the idealized model results the flow through the contralateral PCoA is much less
affected, even with the 70% RICA or BA stenoses.

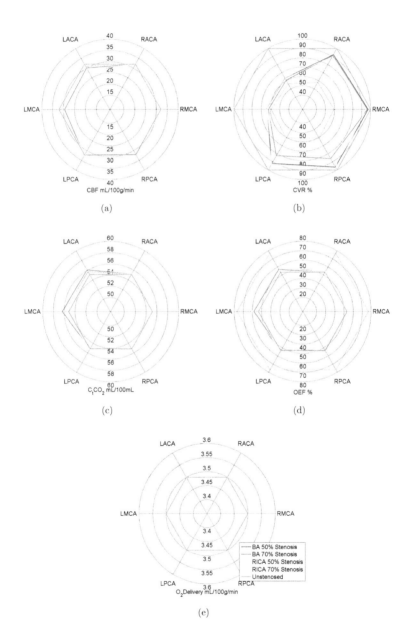

Figure 9.7: The end response to an occlusion of the LICA in combination with other unilateral stenoses for the complete circle of Willis (a) the efferent CBF, (b) the efferent CVR, (c) the efferent C_tCO_2, (d) the efferent OEF, and (d) the efferent O_2 delivery.

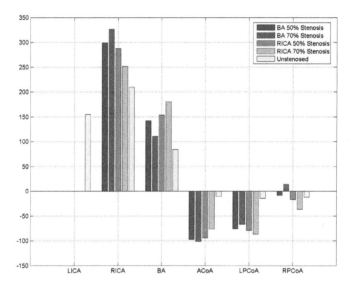

Figure 9.8: The end response to an occlusion of the LICA in combination with other uni-lateral stenoses for the complete circle of Willis illustrating the afferent and communicating artery flowrates.

9.2.2   RICA Occlusion

The results for the autoregulatory response to the RICA occlusion in combination with a unilateral stenosis are presented in Figures 9.9(a) - 9.9(e). It can be observed in Figure 9.9(a) that there is a similar pattern to the LICA occlusion where the CBF is not completely restored in the ipsilateral RACA and RMCA, but in this case the reduction in efferent CBF is so great that CVR for these two arteries reaches its lower limit of autoregulation (Figure 9.9(b)). As a consequence of this the C_tCO_2 in these arteries remains elevated at the end of the simulation (Figure 9.9(c)), but by far the more significant result is that with a 70% LICA stenosis, the OEF for the RMCA reaches its upper limit (and very nearly the RACA too) (Figure 9.9(d)) meaning that O_2 delivery cannot be maintained to the cerebral territory perfused by the RMCA (Figure 9.9(e)).

The afferent and communicating artery flowrates (Figure 9.10) show that the ACoA flowrates are of similar magnitude as with the LICA occlusion simulations, although in this case the direction of flow is from left to right, rerouted to perfuse the RACA and RMCA. Furthermore, it can be observed that the flow through the ipsilateral RPCoA is of a similar magnitude as through the LPCoA with a LICA occlusion, and the contralateral LPCoA does not show an appreciable increase in collateral flow, except for the case of the 70% LICA stenosis, in which case almost as much collateral flow is provided as through the RPCoA. It is interesting that in the case of a LICA occlusion and a 70% RICA stenosis, O_2 delivery can be maintained, whereas with a RICA occlusion and a 70% LICA stenosis it cannot. This result would most likely not be reproduced if the same scenario were imposed on the idealized circle of Willis geometry, and the reason that it occurs is that under normal conditions there is approximately 20mL/min more through the RMCA than the LMCA and 10mL/min more through the RACA than the LACA, meaning that more collateral flow would need to be provided to the right anterior of the CoW. Since the ipsilateral RPCoA is unable to provide this collateral flow, the end reduction in blood flow is so great that even maximizing OEF is not enough to restore O_2 delivery.

Figures 9.11(a) and 9.11(b) provide a visualization of the pressure and velocity field for the case of the RICA occlusion and a 70% LICA stenosis. The main features worthy of note are the reduction in blood pressure caused by both the occlusion and the stenosis, as well as the secondary and helicoidal flows which occur immediately distal to the stenosis and distal to the ACoA respectively, as blood flow is rerouted through the ACoA to supply the RMCA. It is important to note that these secondary flows would be exaggerated and very different if the simulation was pulsatile due to the systolic deceleration, although due to the large differences in timescale between the cardiac pulse and the time of the simulation, they won't affect the overall result. The important point however is that these secondary flow features are highly 3D and could not be captured properly by a 1D modelling approach, highlighting the importance of 3D modelling.

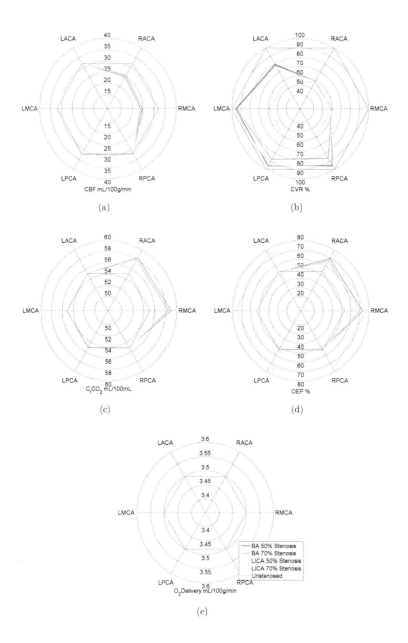

Figure 9.9: The end response to an occlusion of the RICA in combination with other unilateral stenoses for the complete circle of Willis (a) the efferent CBF, (b) the efferent CVR, (c) the efferent C_tCO_2, (d) the efferent OEF, and (d) the efferent O_2 delivery.

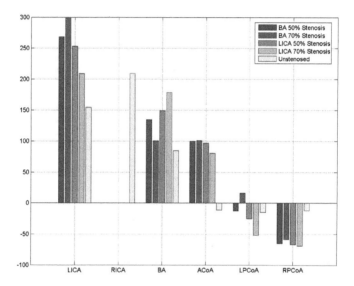

Figure 9.10: The end response to an occlusion of the RICA in combination with other unilateral stenoses for the complete circle of Willis illustrating the afferent and communicating artery flowrates.

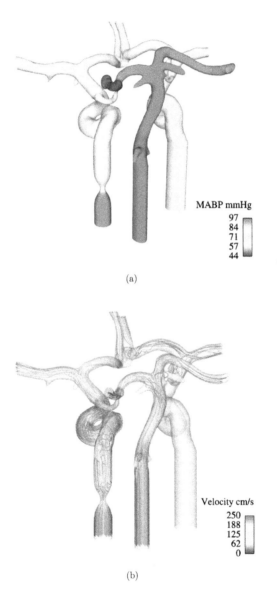

(a)

(b)

Figure 9.11: The end response to an occlusion of the RICA in combination with a 70% LICA stenosis for the complete circle of Willis model, illustrating (a) MABP at the arterial wall and (b) a streamline plot colored by the blood velocity.

9.2.3 BA Occlusion

The results for the autoregulatory response to the BA occlusion in combination with a unilateral stenosis are presented in Figures 9.12(a) - 9.12(e). It can be observed in Figure 9.12(a) that CBF is maintained throughout the CoW and hence C_tCO_2 remains at its normal level at the end of the simulation (Figure 9.12(c)). There is a substantial reduction in CVR in both PCA's in order to restore CBF (Figure 9.12(b)) but the reduction is only approximately half of the lower limit of autoregulation. An interesting result is that the combination of a posterior afferent occlusion with an anterior afferent stenosis in either ICA causes very little effect and a much smaller reduction of about 15% compared to a reduction of approximately 25% for the case of an anterior afferent stenosis and anterior afferent occlusion. Since CBF is maintained in all efferent arteries, OEF remains at its normal level (Figure 9.12(d)) and O_2delivery is maintained (Figure 9.12(e)).

The afferent and communicating artery flowrates (Figure 9.13) show that as would be expected, both PCoA flowrates reverse direction compared to normal conditions, but the amount of collateral flow provided by each is less than in the previous two cases with the ICA occlusions. The reason for this reduced amount of collateral flow is that the average blood flow through the PCA's is only approximately 29mL/min which is significantly less than the average of approximately 70mL/min and 125mL/min through the ACA's and MCA's respectively and hence this small amount of collateral flow can be provided by the two PCoA's. The direction of the ACoA flowrate is from right to left for the cases containing a LICA stenosis, but an interesting result is that even when there is a 50% RICA stenosis, the flow is still essentially of the same magnitude as under normal conditions and directed from right to left. It requires a 70% stenosis in the RICA to cause a large enough reduction in perfusion pressure before collateral flow is provided from the left side of the CoW

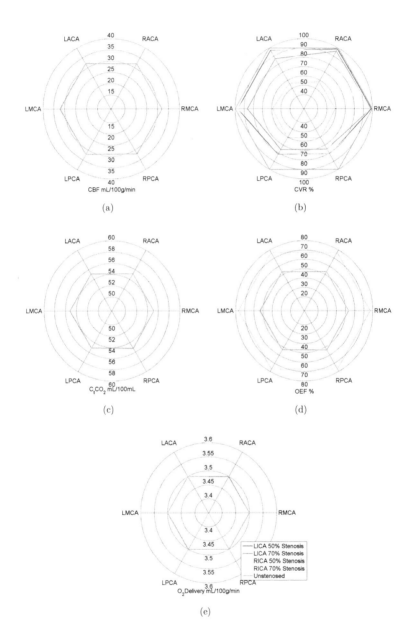

Figure 9.12: The end response to an occlusion of the BA in combination with other unilateral stenoses for the complete circle of Willis (a) the efferent CBF, (b) the efferent CVR, (c) the efferent C_tCO_2, (d) the efferent OEF, and (d) the efferent O_2 delivery.

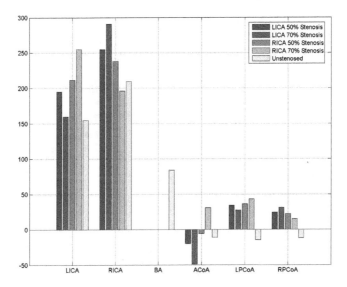

Figure 9.13: The end response to an occlusion of the BA in combination with other unilateral stenoses for the complete circle of Willis illustrating the afferent and communicating artery flowrates.

9.3   Missing Posterior Communicating Artery

The results of the PC MRI scans in terms of the average efferent artery flowrates for the missing PCoA circle of Willis are presented in Table 9.2 along with the corresponding values of CVR when the model was in its set point state. The total efferent blood flow in this case is 657.6mL/min which is significantly higher compared to the complete CoW and closer to the generally accepted value. As illustrated in Figures 9.14(c) and 9.14(d) the higher amount of blood flow under normal conditions results in set point levels of carbon dioxide and oxygen extraction in the tissue that are approximately 51.5mL/100mL and 36% respectively, which are both lower than the case of the complete CoW.

Table 9.2: The efferent flowrates for the missing PCoA circle of Willis measured with PC MRI, and the values of CVR required to achieve them in the numerical simulation.

	LACA	RACA	LMCA	RMCA	LPCA	RPCA
CBF mL/min	126.2	101.8	113.6	157.0	74.5	84.5
CVR mm^{-2}	1232	1136	1140	750	1028	1606

9.3.1   LICA Occlusion

The results for the autoregulatory response to the LICA occlusion in combination with a unilateral stenosis are presented in Figures 9.14(a) - 9.14(e). It can be observed in Figure 9.14(a) that for the BA stenoses and the RICA 50% stenosis, CBF is completely restored in all efferent arteries, but for the RICA 70% stenosis there is a reduction in efferent CBF over the entire anterior portion of the CoW (meaning both ACA's and MCA's). This result is different compared to the complete circle of Willis, where the major reduction in CBF occurred only on the occluded side. It can also be observed in Figure 9.14(b) that for most simulations the reduction in CVR is shared more equally between the left and right sides of the CoW, compared to the LICA occlusion of the complete CoW where the ipsilateral anterior arteries were more severely affected. For the case of the RICA 70% stenosis, CVR is essentially at the lower limit of autoregulation over the entire anterior portion of the CoW, and as a result C_tCO_2 remains elevated in these arteries (Figure 9.14(c)) and so too does OEF (Figure 9.14(d)), although not approaching the upper limit, so O_2 delivery is maintained to every territory (Figure 9.14(e)).

The afferent and communicating artery flowrates (Figure 9.15) show that under normal conditions the flowrate through the LPCoA and the ACoA are of a similar magnitude of approximately 25mL/min, which is slightly higher than for the complete CoW. The direction of the LPCoA flow is negative, indicating that the afferent supply from the BA perfuses the anterior of the CoW, and that in the ACoA is negative, indicating that the afferent supply from the RICA perfuses the LACA. One of the more interesting results is that generally there is almost twice as much flow through the ACoA compared to the complete CoW in response to the stenosis and occlusion, although flow through the ipsilateral LPCoA is of a similar magnitude. The reason for the increased collateral flow through the ACoA arises from the fact that the diameter of the ACoA for the complete CoW is only approximately 1.2mm, whereas the ACoA diameter for this missing PCoA circle of Willis is approximately 2mm. When a greater amount of collateral flow can be provided through a communicating artery then blood supply which would normally perfuse a given region of the brain, is in effect 'stolen' from that region to be rerouted around the CoW. So the greater amount of ACoA collateral flow provides the reason for the much larger effect of the stenosis and occlusion on the anterior flowrates compared to the same scenario with the complete CoW. It can also be observed that even when there is a 70% BA stenosis, the flow through the ipsilateral LPCoA is still directed toward the occluded LICA. This means that the effect of this stenosis must be overcome entirely by the reduction in CVR of the PCA's since no collateral flow is able to be provided to the posterior of the CoW. The afferent flowrate results show that the approximate 190mL/min normally supplied by the LICA is made up for mainly through a large increase in afferent supply through the RICA, but even in the case where it is heavily stenosed (i.e. the 70% RICA stenosis simulation) the increase in afferent supply is similar to that through a completely unstenosed BA.

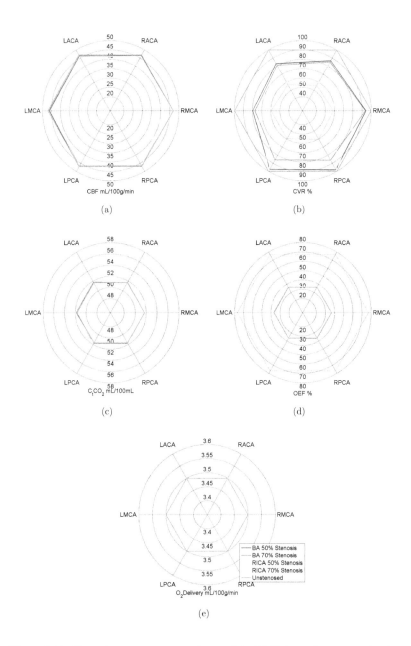

Figure 9.14: The end response to an occlusion of the LICA in combination with other unilateral stenoses for the missing PCoA circle of Willis (a) the efferent CBF, (b) the efferent CVR, (c) the efferent C_tCO_2, (d) the efferent OEF, and (d) the efferent O_2 delivery.

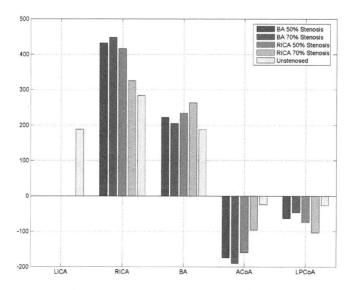

Figure 9.15: The end response to an occlusion of the LICA in combination with other unilateral stenoses for the missing PCoA circle of Willis illustrating the afferent and communicating artery flowrates.

9.3.2   RICA Occlusion

The results for the autoregulatory response to the RICA occlusion in combination with a unilateral stenosis are presented in Figures 9.16(a) - 9.16(e). It can be observed that there is a similar pattern compared to the LICA occlusion where CBF is not completely restored for most simulations throughout the anterior of the CoW (although given more simulation time they may have been restored) and with the LICA 70% stenosis there is a large reduction in efferent CBF. The CVR of these arteries approaches the lower limit of autoregulation and for the case of the 70% LICA stenosis reaches it, (Figure 9.16(b)) and as a result C_tCO_2 remains elevated on in these arteries (Figure 9.16(c)) and OEF remains elevated (Figure 9.16(d)), but does not reach its upper limit, hence O_2 delivery is maintained (Figure 9.16(e)). The interesting result in this case is that with the occlusion of the RICA there is only one pathway for collateral flow, namely through the ACoA, which supplies both the RACA and the RMCA. For the complete CoW, it was shown that given this combination of occlusion and stenosis, O_2 delivery could not be maintained even with the presence of two collateral flow pathways to the right anterior of the CoW. For this geometry it can be observed that the larger ACoA is more effective at providing collateral flow than the case the complete CoW where there was a smaller ACoA with an RPCoA present.

The afferent and communicating artery flowrates (Figure 9.17) show that there is a large amount of collateral flow rerouted through the ACoA, which reverses direction from the normal case, providing the blood supply to the right side of the CoW. It is also interesting to note that there is significantly more collateral flow compared to the previous LICA occlusion simulations (a maximum of approximately 250mL/min for this case compared to a maximum of approximately 190mL/min for the LICA occlusion). This is due to the greater amount of blood flow through the RACA and RMCA compared to the LACA and LMCA under normal conditions, as well as the lack of an RPCoA to contribute to the collateral flow. It can also be observed that the collateral flow through the LPCoA is of similar magnitude compared to the case of the LICA occlusion, further emphasizing the importance of the ACoA for this geometry to reroute the large amount of flow provided almost entirely by an increase in afferent supply from the LICA.

Figures 9.18(a) and 9.18(b) provide a visualization of the pressure and velocity field for the case of the RICA occlusion and a 70% LICA stenosis. Again it can be observed that there is a large reduction in blood pressure caused by both the occlusion and the stenosis, as well as large secondary and helicoidal flows which occur immediately distal to the stenosis and distal to the ACoA respectively, as blood flow is rerouted through the ACoA to supply the RMCA.

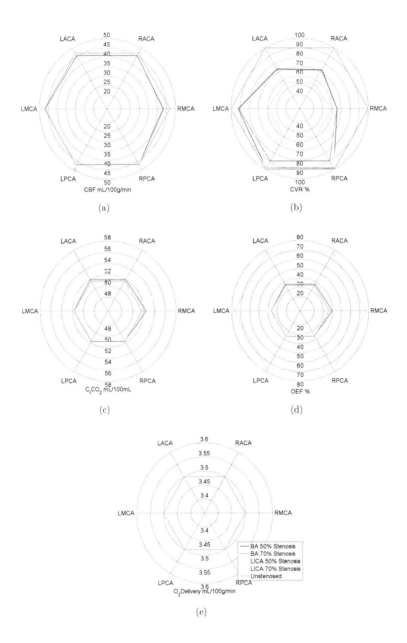

Figure 9.16: The end response to an occlusion of the RICA in combination with other unilateral stenoses for the missing PCoA circle of Willis (a) the efferent CBF, (b) the efferent CVR, (c) the efferent C_tCO_2, (d) the efferent OEF, and (d) the efferent O_2 delivery.

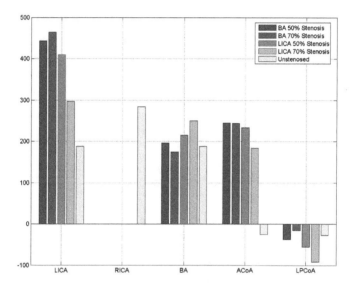

Figure 9.17: The end response to an occlusion of the RICA in combination with other unilateral stenoses for the missing PCoA circle of Willis illustrating the afferent and communicating artery flowrates.

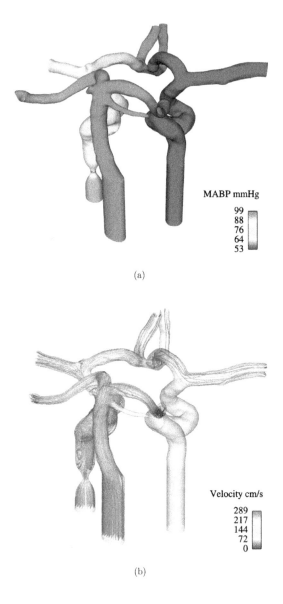

MABP mmHg

99
88
76
64
53

(a)

Velocity cm/s

289
217
144
72
0

(b)

Figure 9.18: The end response to an occlusion of the RICA in combination with a 70% LICA stenosis for the missing PCoA circle of Willis model, illustrating (a) MABP at the arterial wall and (b) a streamline plot colored by the blood velocity.

9.3.3   BA Occlusion

The results for the autoregulatory response to the BA occlusion in combination with
a unilateral stenosis are presented in Figures 9.19(a) - 9.19(e). It can be observed
in Figure 9.19(a) that CBF is maintained throughout the entire anterior of the
CoW, but not maintained in the PCA's for any stenosed simulation. The effects of
stenosis cause a minor reduction in CVR in the anterior efferent arteries (Figure
9.19(b)), while reaching the lower limit of autoregulation in the PCA's. As a result
there is an increase in C_tCO_2 in the PCA's (Figure 9.19(c)) and an increase in
OEF (Figure 9.19(d)), which approaches the upper limit but does not in fact reach
it, hence O_2 delivery is maintained to all territories of the CoW (Figure 9.19(e)).
This is another interesting result because similar to the last occlusion case, only
one pathway for collateral flow is provided, and furthermore because the LPCoA
for this geometry is of a smaller diameter than the ACoA (approximately 1.5mm in
diameter compared to approximately 2mm for the ACoA), and is obviously much
longer in length, therefore its capacity for providing collateral flow is much less, as
has been illustrated with the first two occlusion cases. It can be observed that there
is a significant reduction in efferent CBF in the PCA's at the end of the simulation,
and hence the LPCoA cannot provide the required collateral flow. However, an
important point however is that because this circle of Willis had a greater amount
of blood flow and a correspondingly lower OEF under normal conditions, then
the capacity for OEF to increase is greater compared to the complete CoW. To
emphasize the point, the complete CoW could increase OEF by approximately 23%
before reaching the upper limit, while the missing PCoA can increase OEF by
approximately 41%, meaning that although the PCA flow is severely reduced in this
case, O_2 delivery is able to be maintained.

The afferent and communicating artery flowrates (Figure 9.20) show that for
the BA occlusion, the ACoA is much less important, generally providing a smaller
amount of collateral flow compared to the ICA occlusions, which is directed towards
the stenosed artery (either the LICA or the RICA) although it can be seen that
with a 50% RICA stenosis, the effect of the stenosis is small enough that the ACoA
flowrate barely changes from its normal value. Interestingly, while the LPCoA flow is
directed from anterior to posterior in this case, the amount of flow is not significantly
greater than for the ICA occlusion cases, indicating that the limit of collateral flow
through this LPCoA is approximately 100mL/min. The afferent flowrate results
show that the increase in afferent supply comes largely through the unstenosed

artery since the larger ACoA allows for more afferent supply through the RICA in the scenario of a severe LICA stenosis. In addition, it can reroute blood through the ACoA and LPCoA to supply the PCA's, rather than draw more supply through the stenosed LICA.

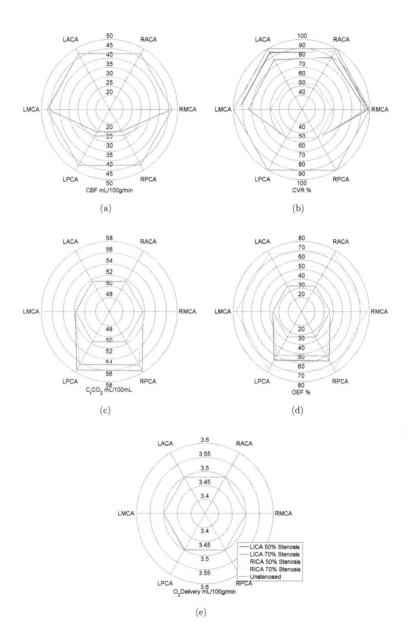

Figure 9.19: The end response to an occlusion of the BA in combination with other unilateral stenoses for the missing PCoA circle of Willis (a) the efferent CBF, (b) the efferent CVR, (c) the efferent C_tCO_2, (d) the efferent OEF, and (d) the efferent O_2 delivery.

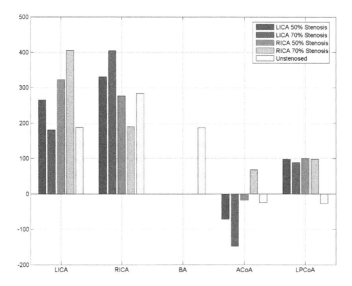

Figure 9.20: The end response to an occlusion of the BA in combination with other unilateral stenoses for the missing PCoA circle of Willis illustrating the afferent and communicating artery flowrates.

9.4   Fused Anterior Cerebral Arteries

The results of the PC MRI scans in terms of the average efferent artery flowrates for
the fused ACA circle of Willis are presented in Table 9.3 along with the corresponding values of CVR when the model was in its set point state. The total efferent
blood flow in this case is 853.1mL/min which is the highest measured value of all
the volunteers and close to the generally accepted average value of 750mL/min. As
illustrated in Figures 9.21(b) and 9.21(d) the higher amount of blood flow under
normal conditions results in the lowest normal levels of carbon dioxide and oxygen
extraction in the tissue, which are approximately 50mL/100mL and 28% respectively.

Table 9.3: The efferent flowrates for the fused ACA circle of Willis measured with PC
MRI, and the values of CVR required to achieve them in the numerical simulation.

	LACA	RACA	LMCA	RMCA	LPCA	RPCA
CBF mL/min	117.6	144.8	171.7	201.0	111.5	106.5
CVR mm^{-2}	631	504	550	480	829	875

9.4.1   LICA Occlusion

The results for the autoregulatory response to the LICA occlusion in combination
with a unilateral stenosis are presented in Figures 9.21(a) - 9.21(e). It can be
observed in Figure 9.21(a) that similar to the missing PCoA circle of Willis, the
effects of an occlusion and stenosis have a greater effect in the entire anterior of
the CoW compared to the complete CoW model. In contrast to the missing PCoA
circle of Willis however, while it took a 70% RICA stenosis to result in the efferent
CBF not being able to be restored to normal levels, this can be seen to occur with
all stenoses, even the BA stenosis simulations. Also, in contrast to the previous two
CoW models, it can be seen that a 70% BA stenosis results in efferent CBF to
the PCA's not being able to be restored. It can be seen that CVR in the LMCA
reaches its lower limit and is approached in both of the ACA's (Figure 9.21(c)) and
as a result C_tCO_2 remains elevated around the anterior of the CoW (Figure 9.21(b)),
with a corresponding increase in OEF (Figure 9.21(d)),though the upper limit is not
achieved in any efferent artery, hence O_2 delivery is maintained everywhere (Figure
9.21(e)).

Since there exists no ACoA for this geometry, but the fusion of the ACA's is not broken, then collateral flow can still be provided around the anterior of the CoW. In order to observe this effect Figure 9.22 exhibits the flowrates through the A1 segments of the LACA and the RACA. Under normal conditions it can be observed that there is very little flow through either of the PCoA's with only approximately 5mL/min through the LPCoA and 10mL/min through the RPCoA, both of which are directed from posterior to anterior. Considering the anterior of the CoW the opposing directions of flow for both A1 segments illustrate that they are both directed toward the fusion and contributing blood supply to the distal LACA and RACA, although the slightly higher flowrate through the $LACA_1$ indicates that the LICA is contributing slightly more of this supply than the RICA. One of the more interesting results of these simulations is that in response to any combination of stenosis with the LICA occlusion the flow through the $LACA_1$ reverses direction to supply the LMCA. While it is difficult to quantify the amount of collateral flow provided, if the definition was based on the difference between the $LACA_1$ flowrate under normal conditions and in response to the stenosis, then the collateral flow through the anterior of the CoW would be approximately 250mL/min, the largest amount for the three CoW geometries investigated and is accompanied by a large increase in flow through the $RACA_1$. It is interesting to note that the communicating arteries are present in order to provide the capacity for collateral flow, although their smaller diameters compared to the other arteries of the CoW, and hence their higher resistance, limits their capacity for collateral flow. In this particular model the absence of the ACoA means that the resistance to collateral flow is effectively based on the resistance of the ACA_1 segments and since the average diameter for these arteries is approximately 2.5mm, there is correspondingly a much greater capacity for collateral flow compared to the other two models. It is also interesting to note that due to a much greater amount of efferent CBF under normal conditions, then the increased capacity for collateral flow around the anterior of the CoW is still not enough to allow for the complete restoration of CBF in the anterior efferent arteries for most simulations. Another interesting result is that the flow through the PCoA's barely changes due to the presence of the occlusion and stenosis. This can be understood by considering that the diameters of both PCoA's is only approximately 0.7mm and hence there is very little capacity for collateral flow. This is also why the BA stenosis had such an effect on the PCA flowrates and a corresponding reduction in CVR required to restore CBF to its normal value (or not restore as would be the case for the 70% BA stenosis).

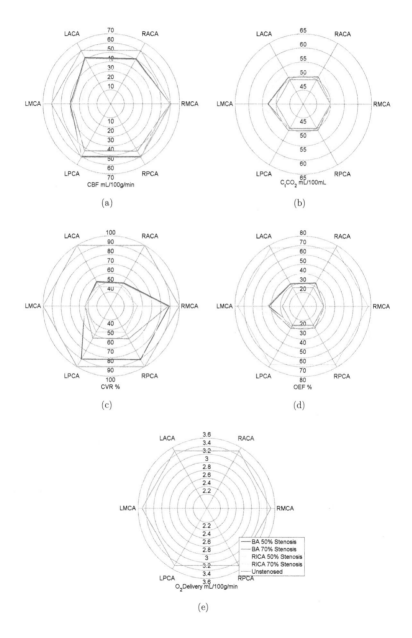

Figure 9.21: The end response to an occlusion of the LICA in combination with other unilateral stenoses for the fused ACA circle of Willis (a) the efferent CBF, (b) the efferent CVR, (c) the efferent C_tCO_2, (d) the efferent OEF, and (d) the efferent O_2 delivery.

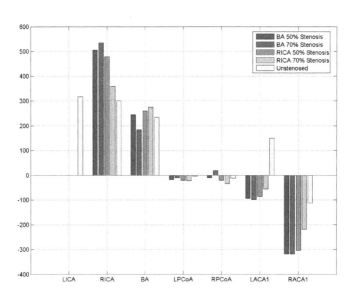

Figure 9.22: The end response to an occlusion of the LICA in combination with other unilateral stenoses for the fused ACA circle of Willis illustrating the afferent and communicating artery flowrates.

9.4.2   RICA Occlusion

The results for the autoregulatory response to the RICA occlusion in combination with a unilateral stenosis are presented in Figures 9.23(a) - 9.23(e). It can be observed there is almost a symmetric pattern compared to the LICA occlusion in terms of the effect on CBF (Figure 9.23(a)). Despite the fact that there is more blood flow through the RMCA compared to the LMCA under normal conditions, the reduction in efferent CBF from the RICA occlusion is less compared to the reduction in LMCA flow in response to the LICA occlusion. It can be observed that CVR around the entire anterior of the CoW has either reached or approached the lower limit of autoregulation (Figure 9.23(a)) and as a result C_tCO_2 remains elevated around the anterior of the CoW (Figure 9.23(b)) and OEF remains elevated, although not reaching the upper limit so that O_2 delivery is maintained everywhere (Figure 9.23(e)).

The afferent and communicating artery flowrates (Figure 9.24) show that the increase in collateral flow through the RPCoA is slightly more significant in this case compared to the LICA occlusion, given that it is now ipsilateral to the occlusion and hence has a much greater pressure gradient across it. The amount of flow is still much smaller than for the other two CoW models. In terms of collateral flow around the anterior of the CoW, using the same definition as for the LICA occlusion, the amount of collateral flow is generally lower, at only approximately 200mL/min being rerouted from the left to the right to perfuse the RMCA, but still high compared to the other CoW models.

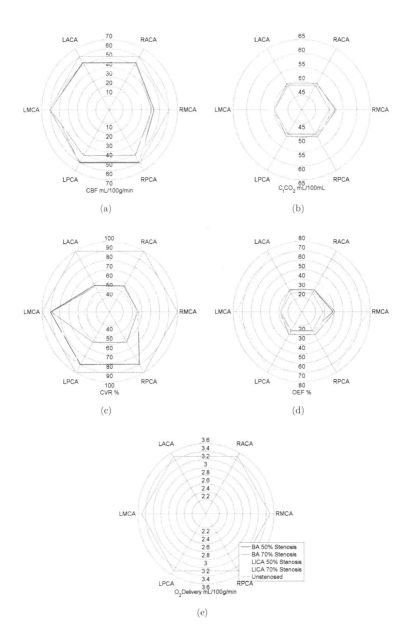

Figure 9.23: The end response to an occlusion of the RICA in combination with other unilateral stenoses for the fused ACA circle of Willis (a) the efferent CBF, (b) the efferent CVR, (c) the efferent C_tCO_2, (d) the efferent OEF, and (d) the efferent O_2 delivery.

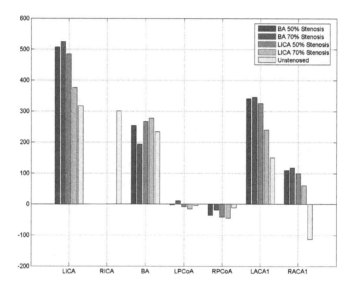

Figure 9.24: The end response to an occlusion of the RICA in combination with other unilateral stenoses for the fused ACA circle of Willis illustrating the afferent and communicating artery flowrates.

9.4.3   BA Occlusion

The results for the autoregulatory response to the BA occlusion in combination with a unilateral stenosis are presented in Figures 9.25(a) - 9.25(e). In this case the inability of the PCoA's to provide collateral flow becomes very important as it can be observed that for all simulations the efferent CBF in both PCA's cannot be restored (Figure 9.25(a)) and furthermore its reduction is very large. Conversely, the stenosis in either ICA does not cause a significantly large reduction in afferent supply and as a result CBF to the anterior part of the CoW is maintained. The resulting CVR for the PCA's reaches the lower limit of autoregulation for all simulations (Figure 9.25(c)) and as a result C_tCO_2 remains largely elevated in these arteries (Figure 9.25(b)) and so too does OEF (Figure 9.25(d)). Most important is the fact that OEF for the PCA's reaches its upper limit and hence the reduction in efferent blood flow to the PCA's is so great that O_2 delivery cannot be maintained to them (Figure 9.25(e)). It is interesting to note that because this circle of Willis had the highest CBF under normal conditions, then the OEF was lowest and the capacity to increase was the highest for all three models, in that it could increase by 49%, although this increased capacity proved to be still not enough to maintain O_2 delivery. This result is remarkable in that it would generally be expected that the presence of two collateral pathways to the posterior of the CoW would be more effective for rerouting blood flow than one collateral pathway, as was the case for the missing PCoA circle of Willis. It can be observed however that because the single PCoA was more than twice the diameter (i.e 1.5mm compared to approximately 0.7mm for the PCoA's of the fused ACA circle of Willis), it was able to provide significantly more collateral flow, which coupled with the fact that there was an average of approximately 28mL/min less through each PCA under normal conditions, means that O_2 delivery could be maintained in the case of a BA occlusion. This simulation illustrates a pathological condition which would very clearly not be survivable for the individual possessing this circle of Willis.

The afferent and communicating artery flowrates (Figure 9.26) show that even with the large pressure gradient between the PCoA's, caused by the BA occlusion, the flowrates through them are not significantly increased compared to the other occlusion simulations, or even compared to that under normal conditions, indicating that their limit for providing collateral flow is significantly smaller than the other CoW models. It can be seen that in the anterior of the CoW there is a significant amount of collateral flow being rerouted towards the stenosed artery, although the amount is significantly less than in the case of an ICA occlusion in combination with an ICA stenosis.

Figures 9.27(a) and 9.27(b) provide a visualization of the pressure and velocity field for the case of the BA occlusion and a 70% RICA stenosis. Again it can be observed that there is a large reduction in blood pressure caused by both the occlusion and the stenosis, as well as the large secondary and helicoidal flows which occur immediately distal to the stenosis and as blood flow is rerouted through the A1 segments of the ACA's.

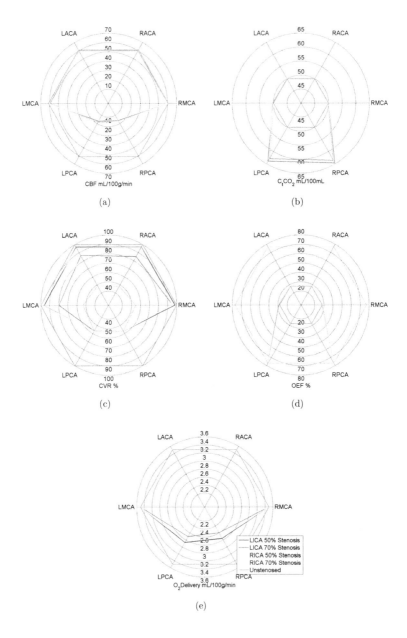

Figure 9.25: The end response to an occlusion of the BA in combination with other unilateral stenoses for the fused ACA circle of Willis (a) the efferent CBF, (b) the efferent CVR, (c) the efferent C_tCO_2, (d) the efferent OEF, and (d) the efferent O_2 delivery.

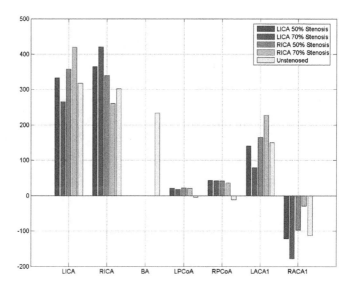

Figure 9.26: The end response to an occlusion of the BA in combination with other unilateral stenoses for the fused ACA circle of Willis illustrating the afferent and communicating artery flowrates.

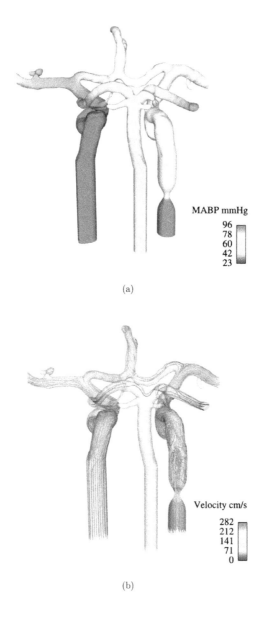

(a)

(b)

Figure 9.27: The end response to an occlusion of the BA in combination with a 70% RICA stenosis for the fused ACA circle of Willis model, illustrating (a) MABP at the arterial wall and (b) a streamline plot colored by the blood velocity.

The results of these patient specific simulations provide some interesting and unexpected results. The major finding is that collateral flow patterns around the CoW and the overall response to a stenosis or occlusion are highly patient specific. While the initial hypothesis with the idealized models was that the anatomical configuration of the circle of Willis would be the major factor determining the ability to provide collateral flow and hence maintain oxygen delivery, these results show that it is in fact more complicated that this, depending upon the diameters of the communicating arteries as well as the amount of efferent blood flow under normal conditions. While it is obvious that the capacity to provide collateral flow depends upon the diameter of the vessel, the challenge is predicting exactly how much collateral flow will be provided in response to an occlusion or stenosis and furthermore, predicting whether or not it will be enough collateral flow to maintain O_2 delivery. It has been shown that there are situations where the absence of a communicating artery is of no detriment to collateral flow as the remaining communicating arteries are larger, and conversely there are situations where all of the communicating arteries are present, but too small to provide adequate collateral flow. These situations emphasize the importance of utilizing a modelling technique of this sort to be able to accurately predict, on a patient specific basis, the ability of a particular circle of Willis to provide collateral flow. Furthermore, the strong secondary and helicoidal flows present when there is a stenosis in an artery and also when flow is rerouted through the communicating arteries are highly 3D effects and would not generally be captured by a 1D model of cerebral blood flow. It remains to be seen how the losses in perfusion pressure from the secondary flow affect collateral flow when compared to a 1D model, but these results emphasize the deviation from axisymmetric flow, the basis for a number of 1D models, emphasizing the importance of 3D modelling.

At present, the major limitations to the modelling technique are related to the limits of CVR and OEF. In particular, it has been assumed that the normal or set point levels for CVR occurs when the normal levels of efferent CBF are reached. While this may be a valid assumption for the young, normotensive volunteers (with a very low probability of having significant atherosclerotic plaque formation) selected for this study, if the modelling technique were used clinically, then it may well be used on people that do have significant atherosclerosis. In this case their CVR in some cerebral territories may already be lowered, even approaching the lower limit of autoregulation. Furthermore their OEF may be increased, even approaching its upper limit within a region of the brain. In such a case, the assumption that all efferent CVR and OEF values would be at their set point when the measured efferent flowrates were achieved would be incorrect. What is necessary is an investigation, or incorporation of data from previous investigations, to devise a means by which the estimated reduction in CVR and increase in OEF from their set points could be implemented into the model, such that the effects of further reduction in cerebral blood flow could be accurately simulated.

Chapter 10

Conclusions and Future Work

The simulations presented in this study highlight the fact that cerebral hemodynamics are a very patient specific phenomena. The ability of the brain to maintain oxygen delivery in response to an occlusion and stenosis in an afferent artery was shown to depend upon the particular anatomical configuration of the circle of Willis, although the issue is more complicated than simply the configuration, as the ability to provide collateral flow through the communicating arteries depends greatly upon their diameter in addition to simply their presence. The normal levels of cerebral blood flow were shown to differ significantly between the three patient specific models investigated, and this has important consequences in terms of maintaining oxygen delivery. Firstly, having a lower level of cerebral blood flow entails that a lesser flow would be necessary through the communicating arteries, or a stenosed region, in order to restore blood flow to the normal levels. Another important consequence of this modelling technique is, assuming the same metabolic rate and brain mass between individuals, that a higher level of cerebral blood flow under normal conditions signifies a lower normal oxygen extraction fraction, hence providing a greater capacity for oxygen extraction when blood flow cannot be completely restored, maintaining oxygen delivery to the tissue. A significant result is that the initial goal of using the simulation results to create general guidelines regarding collateral flow patterns for a specific circle of Willis configuration may in fact be flawed. Of the three simulations performed, some unexpected results were obtained resulting from the particular diameters of the communicating arteries, found to be more important than simply their presence. As an example, it was found that the missing PCoA circle of Willis was able to maintain oxygen delivery following a complete occlusion of the RICA in combination with a 70% LICA stenosis, whereas the complete circle of Willis was not. Intuitively it would seem that in this scenario the absence of the RPCoA to provide collateral flow would be severely detrimental, but the increased diameter of the ACoA allowed a far greater amount of collateral flow to be rerouted around the anterior of the CoW compared to the complete configuration. As a second example, it was found that the fused ACA's circle of Willis was unable to maintain oxygen delivery given any occlusion of the BA while the missing PCoA circle of Willis was. It would again be thought that the presence of only one PCoA to provide collateral flow to the entire posterior of the CoW would be deleterious, but since it was approximately twice the diameter of either PCoA in the fused ACA's circle of Willis, it was able to provide more collateral flow than both PCoA's combined.

The results of these patient specific simulations also highlight a number of interesting flow features such as the reversal of the communicating artery flowrates over the cardiac cycle as well as the recirculation immediately downstream of the stenosed regions and most importantly, the helicoidal flow patterns possible in response to the stenosis when a large amount of collateral flow is provided through the communicating arteries. This last result in particular would not be captured via the use of 1D models of cerebrovascular flow, highlighting the importance of 3D modelling in determining accurate pictures of cerebral blood flow.

The simulations performed incorporated the degrees of stenosis deemed to be significant and of common clinical occurrence. While the results illustrate the potentially dangerous pathological conditions for a particular patient, it is unlikely that the future of this research lies in performing vast numbers of possible combinations of occlusion and stenosis for a patient specific model. Instead it is likely that the technique could be applied to individuals already suffering from a stenosis, although possibly before it poses any symptoms. In such a case the modelling technique could be used to estimate of the severity of an increase in the degree of such a stenosis or the possible effects of a bilateral stenosis occurring. Furthermore there is the potential application of providing presurgical risk assessment information where a significant asymptomatic stenosis is known to exist and its removal requires complete occlusion of that artery while a carotid endarterectomy is performed to remove the atherosclerotic plaques. Another feasible use of the model is to determine how far a hypertensive patient's blood pressure could safely be reduced before lowering cerebral perfusion to a potentially dangerous level. This is a common clinical question and current theory on the treatment for hypertension is to be quite aggressive in the reduction in blood pressure, but not being able to predict the effects on cerebral perfusion limits the ability of the clinician to use an aggressive treatment of antihypertensive drugs. Since hypertension and atherosclerosis are linked, it would be a common finding for a hypertensive patient to have some degree of stenosis in one or more of their afferent arteries, in which case lowering the overall systemic blood pressure could have a more significant effect on flow patterns throughout region of the circle of Willis perfused by the stenosed artery, emphasizing the importance of 3D modelling to predict the effect of antihypertensive drugs on collateral flow.

The modelling technique developed in this study presents only a starting point to modelling and understanding cerebral hemodynamics completely and there is a very large scope for validation of the model before it could become accepted as a clinical tool, and extension of the model to encompass other aspects of neurologic dysfunction. An immediate focus of future work needs to be validating the autoregulation mechanism in terms of its response, as well as the lower limits of cerebrovascular resistance. While monitoring the response of the cerebral vasculature to an acute occlusion of one of the afferent arteries presents practical and ethical limitations, one possibility is to begin testing the model on individuals undergoing a surgical procedure such as a carotid endarterectomy. In this case the predicted ability of the individual's cerebral vasculature to maintain oxygen delivery following the clamping of the carotid artery could be determined and the outcome of the simulation compared to the actual clinical outcome. In addition, an in-vivo experiment which can be done without reaching practical and ethical limits is the stimulation of the autoregulatory response via the administration of controlled doses of oxygen and carbon dioxide. Figure 10.1 presents a sample result of one such test performed where an individual was placed inside an MRI scanner and given controlled, increasing doses of carbon dioxide. As can be observed, the resulting increase in carbon dioxide in the arterial blood causes an increase in carbon dioxide concentration in the cerebral tissue and initiates the autoregulatory response to vasodilate, such that cerebral blood flow can be seen to increase with increasing doses. The increases in arterial carbon dioxide concentration could easily be simulated via the blood chemistry aspect of the autoregulation model, and the resulting simulated increase in blood flow compared to the measured result. This type of experiment would begin to validate the autoregulation model.

Another area of the modelling technique requiring improvement arises from the fact that at present four commercial software packages are required in order to proceed from the raw MRI data to a visualization of the CFD results. There is hence a significant amount of intermediate data storage required between the DICOM files

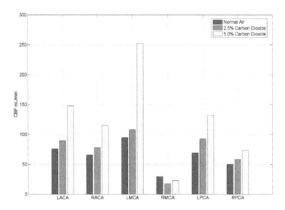

Figure 10.1: A sample clinical experiment aimed at validating the cerebral autoregulation model. In the experiment a volunteer is placed inside an MRI scanner and given controlled doses of carbon dioxide while the cerebral blood flow is measured with phase contrast MRI.

and the post processed result, as well as a significant amount of time to set up the appropriate boundary conditions and solver parameters within the commercial CFD code. Furthermore, both of these issues are compounded by the fact that a separate geometry file was required for every degree of stenosis simulated, which would then be transferred to the mesh generation software and subsequently the CFD code. If this technique was to be implemented as a clinical tool a more streamlined process would be required. Future work should also be aimed at integrating the geometry and grid generation with CFD codes so that simulations can be performed with increased speed via the automated implementation of the boundary conditions, and decreased intermediate data storage, requiring only a single file to store all of the computational grid and simulation results. An additional advantage of a single integrated code would be that alterations in the imposed degree of stenosis in an artery would automatically update through the grid generation process and into the CFD simulation, greatly improving the speed at which simulations could be performed.

An important feature of the modelling technique developed is that it is readily extendible to other regions of the vasculature and there is hence a large scope for moving the afferent terminations further upstream (potentially even to the aortic

Figure 10.2: An example of an extension of the 3D geometry whereby the afferent termination begins at the aortic arch.

arch) and also further downstream of the circle of Willis. Modelling the afferent blood supply from the aortic arch (Figure 10.2) would provide useful information in terms of validating the pressure boundary conditions imposed in the internal carotid and basilar arteries, since the pressure losses and phase difference encountered by blood flow via the four different routes to the circle of Willis could be determined. In this case it may be required to implement a distensible wall in order to accurately represent the hemodynamics, as the larger elastic arteries near the heart exhibit reasonable distensibility. Shifting the afferent terminations would also allow for the simulation and investigation of carotid stenosis in the region of the vasculature where it is actually occurring. Modelling the efferent blood supply through the distal branchings of the anterior, middle and posterior cerebral arteries (Figure 10.3) would provide ample opportunity for future work in that emboli could be tracked through the circle of Willis, down to the smaller arteries where they may in fact block them. Since embolic stroke is by far the most common form of stroke encountered among the general population, then increasing the scope of the modelling technique to investigate this area would be of great clinical value. There is also the potential extension of the model to incorporate 1D models of the vascular network at the efferent 3D terminations. In doing so the lumped parameter resistance of the porous blocks could be abandoned and the alterations in cerebrovascular resistance simulated in the regions where they would actually be occurring. Such an extension would require a considerable increase in the complexity of the CFD simulation, further necessitating the transition from commercial CFD code to an in-house integrated code. One of the major limitations in this study is the resolution of the 1.5T MRI scanner used to generate the patient specific geometries, most limiting in terms of the measurement of the efferent and afferent flowrates. Future work should also include the transition to higher magnetic field strength scanners such as a 3T in order to investigate the accuracy of the efferent blood flow measurements.

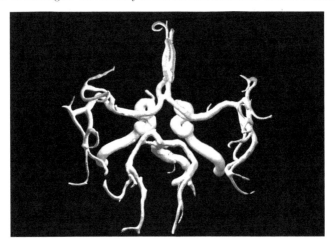

Figure 10.3: An example of an extension of the 3D geometry whereby the efferent terminations extended distally to the circle of Willis.

Finally, as alluded to in Chapter 2 there is a large amount of complexity that could be added to the autoregulation mechanism in terms of the myogenic and shear stress components, as well as an incorporation of more detailed biochemistry simulating brain metabolism. Care must be taken when pursuing this avenue of model improvement however, in that more complex physiological descriptions should include known biochemical pathways with parameters which can be reasonably estimated for humans. Simply generating a mathematical model to capture all of the chemistry can result in a large number of parameters, which may not be determinable and significantly increase the difficulties involved in validation of the autoregulation model.

To summarize, the present study has made a significant contribution to the field of cerebral hemodynamic modelling, integrating the complex 3D patient specific geometries with the relevant physiological control systems present in the cerebral vasculature. However, there remains a large amount of future validatory and model development work which must be performed before the technique can be realized as a clinical tool.

Notes on the Appendices

The modelling of cerebral hemodynamics presented in this thesis incorporates concepts, techniques, and background information from a variety of disciplines, ranging from biochemistry to advanced numerical methods. While it is likely that the reader may be familiar with one or more of these disciplines, it is unlikely that the reader will be familiar with all of them. However, in order to not detract from the focus of the research with large amounts of background information in the body of this thesis, the following four appendices have been compiled by the author to provide a detailed yet concise review of the relevant material. It is the desire of the author that when future work is undertaken in the field of cerebral hemodynamics modelling, these four appendices will provide a much more useful starting point than if the researcher were to begin a new study in order to source all of the information presented here.

Appendix A

Magnetic Resonance Imaging

A.1 Introduction

The purpose of this appendix is to provide an introduction to the technique known as *Magnetic Resonance Imaging (MRI)* to a reader with little or no background in the subject. The information presented here has been compiled by the author from a number of textbooks [56, 66, 100, 132, 143], websites [4, 6], and discussions with experts in the field of MRI, and is designed to provide a detailed yet condensed version of all the background information relevant to the modelling of cerebral hemodynamics. MRI is a technique that was introduced to the field of medical imaging in the 1970's and has widespread clinical use, especially in the imaging of the cerebral vasculature. Its popularity lies in the high contrast that can be generated between the soft tissues, the fact that no ionizing radiation is required, and its flexibility in terms of what can be imaged. This appendix will begin with an introduction to the physics of *nuclear magnetic resonance (NMR)*, continuing with an outline of how the principal is utilized to generate an image. Finally, the application of MRI to the field of *Angiography* will be introduced, as it is the data from two different angiographic techniques which are required to create the computational models.

A.2 Nuclear Magnetic Characteristics of Elements

Protons, neutrons and electrons possess the fundamental properties of angular momentum *(spin)* and electric charge. Since the moving charges gives rise to magnetism, these particles also possess a *magnetic moment vector* μ. Though the neutron is electrically uncharged, charge inhomogeneities on the subnuclear scale result in a magnetic field of opposite direction and approximately the same strength as the proton. A phenomenon known as *pairing* occurs within the nucleus of the atom, where the spins of the constituent protons and neutrons determine the nuclear magnetic moment. If the total number of protons and neutrons in the nucleus is even, the spins cancel out and the magnetic moment is essentially zero. If the total is odd the nuclear spin generates a net magnetic moment. In Nuclear Magnetic Resonance it is unpaired nuclear spins that are of importance.

The key features for biologically relevant elements that are candidates for producing magnetic resonance images include strength of the magnetic moment, the physiologic concentration and the isotopic abundance. Hydrogen, having the largest magnetic moment and the greatest abundance, is by far the best element for general clinical utility. Other elements are orders of magnitude less sensitive when the magnetic moment and the physiologic concentration are considered together. Of these, ^{23}Na and ^{31}P have been used for imaging in limited situations, despite their relatively low sensitivity. Therefore, Hydrogen nuclei (which consists of a single proton) is the principal element used for MRI.

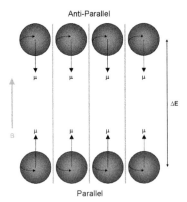

Figure A.1: When placed in an external magnetic field, protons align their spin vector with the field in either a parallel, low energy configuration or an anti-parallel, high energy configuration.

Resonance is an energy coupling that causes the individual nuclei, when placed in a strong external magnetic field, to selectively absorb and later release energy unique

to those nuclei and their surrounding environment. To understand how particles
with spin behave in a magnetic field, consider a proton. When placed in an external
magnetic field the spin vector of the particle aligns itself with the external field
(Figure A.1). There are two possible ways in which this may happen, corresponding
with the two discrete spin states. The first is the low energy configuration where
the spin vector is aligned parallel with the magnetic field and the second is the high
energy configuration where the spin vector is aligned anti-parallel with the magnetic
field. The proton may undergo a transition between the two states by the absorption
of a photon. The frequency of the photon which the proton may absorb is given by:

$$\omega = \gamma B \tag{A.1}$$

where ω is the frequency of the photon and is known as the resonant frequency or
Larmor frequency, B is the magnetic field strength and γ is the *gyromagnetic ratio*,
which is unique for each element and for a proton is 42.58MHz/T. The energy of
the photon E, is given by:

$$E = h\omega \tag{A.2}$$

where h is Plank's constant. In order for a transition to occur the energy of the
photon must be equal to the energy gap ΔE between the parallel and anti-parallel
states. The proportions of spins in each of the energy states obeys Boltzmann
statistics:

$$\frac{N_-}{N_+} = e^{-\frac{E}{kT}} \tag{A.3}$$

where N_- is the number of spins in the high energy state, N_+ is the number of
spins in the low energy state, k is Boltzmann's constant and T is the temperature.
At physiological temperatures and magnetic field strengths used in MRI scanners,
the proportion of spins in the low energy state slightly outnumbers the proportion
in the high energy state, the ratio of excess spins in the low energy states being
approximately 10^{-5}. Considering that there are of the order of 10^{21} protons in a
typical MRI *voxel* volume however, there are approximately 10^{16} more spins in the

low energy state, which is enough to produce an observable magnetic moment.

In addition to the energy separation of the spin states, the protons also experience a torque from the external magnetic field. The combination of their own angular momentum and the effect of the torque to change that angular momentum causes a phenomenon known as *precession* (Figure A.2), where the magnetic moment vectors rotate (precess) about the axis of magnetic field vector at an angular velocity equal to the Larmor frequency described in (A.1).

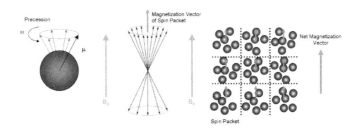

Figure A.2: The net magnetic moment in a spin packet is the vector sum of the spin vectors of all the individual protons in the parallel and anti-parallel directions. Furthermore the net magnetization vector is defined as the vector sum of the magnetization vectors of all the spin packets.

By defining a *spin packet* to be a group of spins experiencing the same magnetic field strength, a macroscopic model of the magnetization can be defined, where the conglomeration of protons precessing in the parallel and antiparallel directions creates a distribution (Figure A.2) with the *magnetization vector* equal to the vector sum of the magnetic moments of all the protons in the spin packet. In addition the *net magnetization vector* $\mathbf{M}$ can be defined as the vector summation of the magnetization vectors from all of the spin packets [66]. At equilibrium, no net magnetization exists perpendicular to the direction of the external magnetic field because the individual protons precess with a random distribution, which effectively cancels out this component. Furthermore, the greater number of spins in the parallel direction means that $\mathbf{M}$ is parallel with the external magnetic field.

A.3 Generation and Detection of the Magnetic Resonance Signal

To aid in the understanding of how the magnetic resonance signal is generated, it is useful to initially define two coordinate systems. The first is the *laboratory frame* of reference (Figure A.3) which is a stationary frame from the observers point of view. By convention the z axis is directed down the bore of the scanner with the main magnetic field $\mathbf{B}_0$. The x and y axes are both perpendicular to the z axis creating a 3D cartesian coordinate system. The protons' magnetic moment precesses about the z axis in a circular geometry in the xy plane. The second coordinate system is the *rotating frame* of reference. The rotating frame is a spinning axis system whereby the x and y axes rotate about the z axis at the precessional frequency of the protons. In this frame, spins appear to be stationary when they rotate at the precessional frequency. Both the laboratory and the rotating frame of reference are useful in explaining various interactions of the protons with externally applied static and rotating magnetic fields.

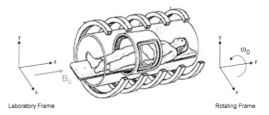

Laboratory Frame   Rotating Frame

Figure A.3: Two frames of reference, useful in MRI. The first is the laboratory frame, which is a fixed 3D cartesian coordinate system with the z axis aligned with the main $\mathbf{B}_0$ field. The second frame of reference is the rotating frame, where the x and y axes rotate about the z axis at the Larmor frequency.

The transient response of the net magnetization vector $\mathbf{M}$ to an applied external magnetic field is governed by the *Bloch* equation:

$$\frac{d\mathbf{M}}{dt} = \gamma \mathbf{M} \times \mathbf{B}_0 - \mathbf{R}\left(\mathbf{M} - \mathbf{M}_0\right) \tag{A.4}$$

where $\mathbf{M}_0$ denotes the net magnetization vector in an equilibrium state and $\mathbf{R}$ is known as a *relaxation matrix* and is given by:

$$\mathbf{R} = \begin{pmatrix} \frac{1}{T_2} & 0 & 0 \\ 0 & \frac{1}{T_2} & 0 \\ 0 & 0 & \frac{1}{T_1} \end{pmatrix} \qquad (A.5)$$

where T_1 and T_2 are known as *relaxation times* and their significance will be discussed shortly. It should be noted however that while the first term on the right hand side of (A.4) can be derived from Newtons second law, the second term is a phenomenological description of the complex microscopic interactions occurring. Since the $\mathbf{B}_0$ field is aligned with the z axis, the equilibrium magnetization will in therefore only have a z component. The net magnetization vector can described by three components and similarly the Bloch equation can be separated into the vector components:

$$\frac{dM_x}{dt} = \gamma \left(M_y B_z - M_z B_y \right) - \frac{M_x}{T_2} \qquad (A.6)$$

$$\frac{dM_y}{dt} = \gamma \left(M_z B_x - M_x B_z \right) - \frac{M_y}{T_2} \qquad (A.7)$$

$$\frac{dM_z}{dt} = \gamma \left(M_x B_y - M_y B_x \right) - \frac{M_z - M_{z0}}{T_1} \qquad (A.8)$$

where M_{z0} is the z component of the equilibrium magnetization vector. The M_z component of the net magnetization is parallel to the applied magnetic field B_0 and is known as *longitudinal magnetization*. It should be apparent that when in equilibrium, $M_z = M_{z0}$, with the amplitude determined by the excess number of protons that are in the low energy state. M_x and M_y are the components of the magnetic moment perpendicular to the applied magnetic field and the magnitude of these two individual components is known as the *transverse magnetization*. At equilibrium, the transverse magnetization is zero, because the vector components of the spins are randomly oriented about the xy plane and cancel each other out. The solution to the Bloch equation yields the x, y and z components of magnetization as a function of time.

The system is perturbed by the application of a pulse of radio frequency electromagnetic radiation (an *RF pulse*) at the precessional frequency (related to the energy gap ΔE), which is absorbed and converts spins from the low energy parallel direction to the higher energy antiparallel direction. Furthermore, the RF pulse has the effect of causing the protons to precess in phase with one another, known as *phase coherence*. As the perturbed system returns to its equilibrium state (i.e. the solution to the Bloch equation), the magnetic resonance signal is produced. Being an electromagnetic wave, the RF energy contains a sinusoidally oscillating magnetic field in the laboratory frame, known as the B_1 field. In the rotating frame the B_1 field will appear stationary relative to M_z. The B_1 field is applied along either the x or y axis and will create a torque on M_z 'tipping' it into the xy plane. If the frequency of the RF energy is not tuned to the Larmor frequency of the protons then the B_1 field will not appear stationary in the rotating frame of reference and resonance will not occur. The angle θ through which the longitudinal magnetization M_z is tipped into the xy plane (Figure A.4) is known as the *flip angle* and is proportional to the duration of the RF energy pulse. For the purposes of this discussion, flip angles of $90°$ will be considered (i.e. the longitudinal magnetization is converted completely into transverse magnetization), but it should be noted that any flip angle may be chosen depending upon the particular type of scan being performed.

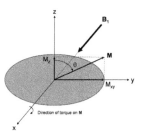

Figure A.4: The RF energy pulse contains an oscillating magnetic field (in the laboratory frame), which creates a torque on the longitudinal magnetization vector, tipping it into the xy plane. The flip angle θ through which the longitudinal magnetization is tipped is proportional to the length of the RF pulse.

Following the RF pulse the transverse magnetization gradually decays to zero and the longitudinal magnetization returns to its equilibrium value. It should be noted that although the total number of spins is always conserved, the net magnetization vector is not always conserved. The solution for the longitudinal component of the Bloch equation is:

$$M_z(t) = M_z(0)e^{\frac{t}{T_1}} + M_{z0}\left(1 - e^{\frac{t}{T_1}}\right) \tag{A.9}$$

where $M_z(0)$ is the initial condition defining the state of the longitudinal magnetization immediately after the RF pulse. As can be observed in (A.8) the return of
M_z to equilibrium is affected by the *spin lattice relaxation time* T_1 (Figure A.5). By
definition T_1 is the time required for the longitudinal magnetization to recover 63%
of its equilibrium value. T_1 relaxation depends on the dissipation of the absorbed
energy into the surrounding molecular lattice and is most efficient when the precessional frequency of the excited protons overlaps the vibrational frequencies of the
molecular lattice.

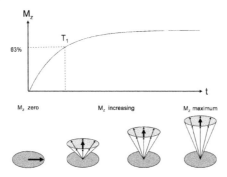

Figure A.5: After being tipped into the xy plane the longitudinal magnetization returns
to its equilibrium value. The spin lattice relaxation time T_1 is the time required for the
longitudinal magnetization to recover 63% of its equilibrium value.

It can be observed in (A.6 and A.7) that the decay of the transverse components
of the magnetization M_x and M_y are affected by the *spin-spin relaxation time* T_2. By
definition T_2 is the time required for the transverse magnetization to decay to 37% of
its maximum value (Figure A.6). This phenomenon is known as *free induction decay*
and occurs because micromagnetic inhomogeneities, intrinsic to the particular tissue
cause the protons to precess at different frequencies corresponding to slight changes
in the local magnetic field strength. As a result there is a loss of phase coherence
between protons and hence a loss of transverse magnetization. T_2 relaxation also
depends upon the molecular structure of the tissue. Mobile molecules exhibit a
relatively long T_2 because rapid molecular motion reduces or cancels the intrinsic

magnetic inhomogeneities. Conversely, as the molecular size increases, constrained molecular motion causes T_2 to shorten.

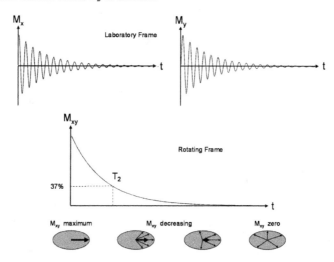

Figure A.6: After being tipped into the xy plane the transverse magnetization rapidly decays due to micromagnetic inhomogeneities within the particular tissue, causing the spins to dephase. In the laboratory frame the x and y transverse magnetization's are decaying sinusoids. In the rotating frame the transverse magnetization follows an exponential decay.

In the presence of *extrinsic* magnetic field inhomogeneities, such as an imperfect $\mathbf{B}_0$ field, the loss of phase coherence occurs more rapidly than from spin-spin interactions alone. The result is to shorten T_2 and the new relaxation time T_2^* is given as:

$$\frac{1}{T_2^*} = \frac{1}{T_2} + \frac{1}{T_2'} \tag{A.10}$$

where T_2' represents the effects of the magnetic field inhomogeneity. The solution to the transverse component of the Bloch equation is:

$$M_{xy}(t) = M_{xy}(0)e^{\frac{t}{T_2^*}}e^{-j\omega_0 t} \tag{A.11}$$

where ω_0 is the Larmor frequency and $M_{xy}(0)$ is the initial transverse magnetization immediately after the RF pulse. The total magnetization in the xy plane M_{xy} is the vector sum of both M_x and M_y, but since these two vectors are 90° out of phase, the total transverse magnetization is represented as a complex quantity:

$$M_{xy} = M_x + jM_y \tag{A.12}$$

When considering all three components of the magnetization vector $\mathbf{M}$ the solution to the Bloch equation describes a helical path in 3D space as $\mathbf{M}$ precesses about the xy plane and gradually returns to the equilibrium magnetization (Figure A.7). In the laboratory frame the transverse magnetization can be seen to oscillate (Figure A.6) and it is this component which is important in terms of generating the MR signal. A receiver coil located in the MRI scanner (which may or may not be the same coil used to generate the RF pules and depends upon the hardware of the particular scanner) will by the principal of Faraday induction, have a voltage induced from the oscillating transverse magnetization. It is this induced voltage that is the signal used to subsequently generate the magnetic resonance image.

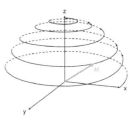

Figure A.7: In the laboratory frame the net magnetization vector follows a helical path in 3D space as the protons return to their equilibrium alignment with the main magnetic field.

It is the difference in both T_1 and T_2 relaxation times for different tissues which allows for the contrast to be achieved when performing a scan of various anatomical components. Table A.1 provides a comparison of both relaxation times for some important biological tissues. It should be apparent that T_1 is always greater than T_2, which is always greater than T_2^* since the loss of phase coherence between protons occurs much more rapidly than the return to equilibrium.

Two important parameters in magnetic resonance imaging are known as the

Table A.1: Relaxation times for important biological tissues [143].

Tissue	T_1 ms (0.5 T)	T_1 ms (1.5 T)	T_2 ms
Fat	210	260	80
Liver	350	500	40
Muscle	550	870	45
White Matter	500	780	90
Grey Matter	650	900	100
Cerebrospinal Fluid	1800	2400	160

Repetition Time (T_R) and the *Echo Time (T_E)*. The repetition time is the period of time between applying subsequent 90° RF pulses. As illustrated in Figure A.5, the longitudinal magnetization takes a period of time to return to its equilibrium value, which is governed by T_1. Usually the next 90° pulse is repeated before a full recovery of the longitudinal magnetization, which means that a partial saturation occurs and the amount of longitudinal magnetization recovered decreases for every subsequent 90° pulse (Figure A.8).

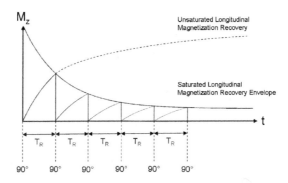

Figure A.8: If the repetition time (the time between subsequent RF pulses) is short enough that the longitudinal magnetization is not allowed to completely return to its equilibrium value, then a saturation of the magnetization can occur, where the amount of magnetization recovered follows a decaying exponential function.

As was illustrated in Figure A.6, following a 90° pulse, where the magnetic moment vector is tipped into the xy plane, the protons rapidly begin to dephase due to T_2^* decay, causing signal loss. The signal loss can be overcome by creating an *echo* in the MR signal. One way to create an echo is by applying a 180° RF pulse at a time $T_E/2$ after the initial 90° pulse and creating what is known as a *spin* *echo* (Figure A.9). The idea behind the technique is that the individual spins in the transverse plane will all dephase at slightly different rates, but when the 180° pulse is applied they will rephase at the same rate at which they dephased, hence at the time T_E after the initial 90° pulse all of the spins in the transverse plane will be back in phase, producing a maximal transverse magnetization signal. It is therefore at this point when the MR signal is acquired. A useful feature of this *spin echo* sequence, is that the effects of inhomogeneities in the external magnetic fields are cancelled out and the peak amplitude of the echo is determined by T_2 instead of T_2^*.

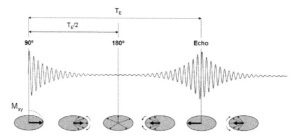

Figure A.9: The actual MR signal measured is an echo, created by rephasing the transverse magnetization. One way to create an echo is to apply a 180° pulse at a time $T_E/2$ creating a spin echo at the time T_E.

It is by choosing the repetition and echo times that the differences in the T_1 and T_2 characteristics of the tissues can be exploited, generating the contrast in the MR image. While there are a number of different choices for the repetition and echo times, each generating a different contrast between tissues and hence a different final MR image, for the purposes of clarity and understanding, one type of contrast will be discussed here, namely a T_1 *weighted image*. A T_1 weighted image generates contrast based primarily on the T_1 characteristics of the tissue, de-emphasizing the T_2 characteristics. This is achieved by choosing a relatively short repetition time, to maximize the differences in longitudinal magnetization during the return to equilibrium, and a short echo time to minimize the T_2 dependency during the signal acquisition. It is the transverse magnetization which is detected to generate the MR image and the actual amount of transverse magnetization (giving rise to the brightness of a particular tissue in the image) depends upon the amount of longitudinal recovery which has occurred in the excited sample. Figure A.10 illustrates the effects of short T_R and T_E on the longitudinal and transverse magnetization, as would occur for T_1 weighted image for some important tissues in the brain. The

amount of longitudinal recovery for the different tissues is projected onto the transverse decay curve to show the amount of transverse magnetization which is available
to generate an MR signal at the time of the echo, which can be extrapolated to show
the intensity in the image of the given tissue.

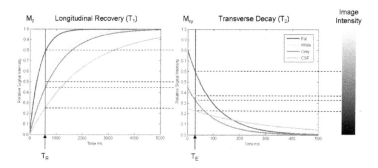

Figure A.10: A T_1 weighted image employs repetition and echo times, maximizing the
effects of the T_1 characteristics of the tissue and minimizing the T_2 characteristics [143].

Figures A.11(a)-A.11(c) illustrate a comparison between different weighting methods for a given slice within a brain. Figure A.11(a) illustrates the T_1 weighted image,
Figure A.11(b) illustrates a T_2 weighted image, where long repetition and echo times
are used to maximize the T_2 characteristics of the tissue and minimize the T_1 characteristics, and Figure A.11(c) illustrates a *proton density* weighted image, where
long repetition times and short echo times are used to maximize the differences in
the number of magnetizable protons per volume of tissue. The main point of illustrating some different weighting techniques is that the same tissue can appear bright
in one type of scan but dark in another and this effect makes MRI useful in clinical
diagnosis.

A.4   Localization of the Magnetic Resonance Signal

In the explanation so far the same $\mathbf{B}_0$ field has been applied to all of the protons
within the MRI scanner. however, in order to obtain a clinically useful image the
magnetic resonance signal must be separable such that the signal corresponding to
a particular spatial location can be determined. The way in which this is achieved is

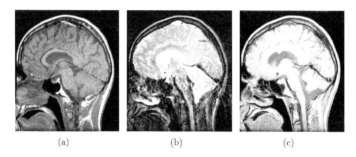

(a)   (b)   (c)

Figure A.11: (a) a T_1 weighted image (b) a T_2 weighted image and (c) a proton density weighted image of a given slice within the brain.

with the use of *magnetic field gradients*. As there are three coordinates required to specify a spatial location, three magnetic field gradients are required (denoted G_x, G_y and G_z respectively) which may be superimposed on the $\mathbf{B}_0$ field. The gradient fields are created by using pairs of coils, which are able to create a linear variation in magnetic field strength in a predefined range (Figure A.12). Although only G_z is shown in the Figure A.12, the concept is the same for the other two coordinates.

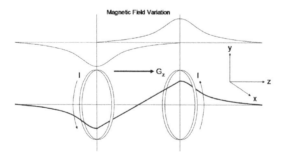

Figure A.12: The linear magnetic field gradients are created using pairs of coils. When an electrical current I is passed through the coil a magnetic field is set up in the space surrounding the coil. By combining coils with opposite polarity a linear magnetic field gradient can be created in a given region inside the MRI scanner.

The first magnetic field gradient is known as the *slice select* (or *slice encode*) gradient and is applied along the z axis in the direction of the $\mathbf{B}_0$ field. Superimposing this gradient field on the $\mathbf{B}_0$ field will cause the Larmor frequency of the protons to vary linearly along the z axis. For a given frequency of the RF pulse, only protons at certain position along the z axis (in an xy plane) will experience resonance (Figure A.13). As can be observed in Figure A.13, the *transmit bandwidth ($\Delta\omega$)* of the RF pulse and the magnitude of the gradient field mean that it is not strictly a plane but rather a thin slab (of *slice thickness* Δz), which is excited. Narrowing the bandwidth and increasing the gradient magnitude will both reduce the thickness of the slab and vice versa.

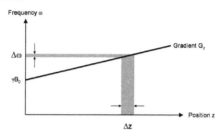

Figure A.13: The magnetic field gradient in the z direction means that the resonant (Larmor) frequency will differ for all protons along the z direction. The application of an RF pulse at a given frequency will therefore only cause resonance and 'select' protons in a given slice along the z axis. The thickness of the slice selected depends upon the strength of the gradient and the transmit bandwidth of the RF pulse.

The second magnetic field gradient is known as the *phase encode* gradient and is applied in either the x or y direction (for the purposes of this discussion it will be assumed that the phase encode gradient is applied in the y direction). During the scanning sequence the phase encode gradient is turned on after the slice select gradient, creating a linear variation in magnetic field strength along the y axis. Within the plane excited by the slice encode gradient the protons will precess at different frequencies due to the variation in magnetic field and will start to become out of phase with one another. After a short duration the phase encode gradient is turned off and the protons will then all precess at the same frequency again, but they will remain out of phase by the same proportion as they were when the phase encode gradient was turned off.

The third magnetic field gradient is known as the *frequency encode* (or *read*) gradient and is applied in the third direction (which for the purposes of this discussion is the x direction). The frequency encode gradient is turned on after the phase encode gradient and is very similar in that it creates a variation in magnetic field strength and causes the protons in the excited slice plane to precess at different frequencies along the x axis. The difference between the frequency and phase encode gradients lies in the fact that the MR signal is obtained while the frequency encode gradient is still turned on. This means that when the MR data is acquired, the protons within the excited slice plane will be precessing at different frequencies in the x direction and will also be out of phase with one another in the y direction (Figure A.14). Processing the MR signal allows for the separation of the various components based on frequency and phase and allows for the signal intensity at different spatial locations to be specified, generating the MR image.

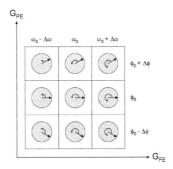

Figure A.14: Nine sample voxels in a selected slice. In the phase encode direction the spins within a voxel develop a phase difference during the time that the phase encode gradient is turned on. This phase difference remains however after the phase encode gradient is turned on. In the frequency encode direction the spins within a voxel precess at different frequency due to the application of the frequency encode gradient. As a result, each voxel in the selected slice has a unique combination of precessional frequency and phase.

There currently exists a multitude of RF pulse and gradient sequences that are used clinically. This discussion does not attempt to cover all of the commonly used sequences, but rather to introduce some of the more important sequencing parameters through a simplified example. Figure A.15 illustrates a simple MRI spin echo sequence showing the timing of the applications of the magnetic field gradients, the RF pulses and the data acquisition. It can be observed that immediately after the slice select gradient is off there is a negative *lobe*, known as a *rewinder*. The reason for the rewinder is that even while the initial 90° pulse is being applied, the spins begin to dephase. The negative lobe will cause the spins to rephase. Furthermore it can be observed that there is a small period of time required for the gradients to reach their required strength which are known as *slew rates*, or *ramp up* and *ramp down* times respectively. These time periods are a hardware limitation as the magnetic field gradients cannot be set up instantaneously.

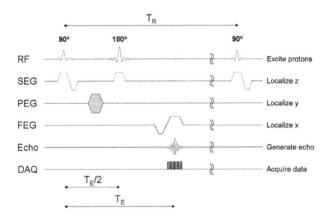

Figure A.15: The pulse and gradient sequence for a typical spin echo. Illustrated are the relative temporal spacings between pulses and applications of the magnetic field gradients.

A.5   Data Acquisition and Image Reconstruction

Understanding how an image is obtained from the magnetic resonance signal requires an introduction to the concept of k *space*, where the raw data is stored and subsequently processed to obtain the final image. A k space plot (Figure A.16(a)) can essentially be thought of as a matrix, the same size as the final MR image (Figure A.16(b)), where each entry contains a complex number. The data contained in k space represents spatial waves of image intensity (units m^{-1}), which may be superimposed to obtain an image. Generally speaking, the low frequency data, stored at the center of k space has the greatest amplitude and provides the contrast in the final image. The high frequency data stored away from the center on the other hand, provides the detail in the final image. The entire scanning process is essentially designed to fill k space with the data which may subsequently processed to obtain the MR image.

While the physical transverse magnetization signal is complex, containing the real magnetization component M_x and the imaginary component M_y, the receiver coil can only in fact detect a real valued voltage. The processing performed in k space requires complex data however, so there is some signal processing which is performed before the raw data is entered into k space. Figure A.17 illustrates the key steps in

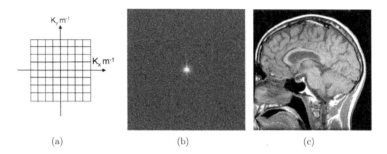

(a)   (b)   (c)

Figure A.16: K space is a complex frequency space where the k space coordinates define spatial waves of various frequency and phase, (a)a k space matrix with the origin at zero frequency and phase (b) a k space image and (c) the resulting MR image.

the signal processing known as *complex demodulation*,which transforms the received signal into a *baseband* signal. The signal received by the coil $S_r(t)$ can be thought of as the real part of the physical signal $S_p(t)$ and is given by $M(t)cos(\omega_0 + \phi(t))$, where $M(t)$ is an amplitude function, ω_0 is the carrier frequency, dependent upon the $\mathbf{B}_0$ field, and $\phi(t)$ is a phase difference brought about by the application of the magnetic field gradients. This signal is split into two parts which are multiplied by a sine and cosine wave, both at the carrier frequency ω_0. Making use of trigonometry identities, the multiplication gives two components for each signal; a high frequency component $2\omega_0 + \phi(t)$ and a low frequency component $\phi(t)$. By passing this signal through a low pass filter the high frequency component can be removed, thereby eliminating the carrier frequency from the signal. The signals then pass through an analog to digital converter and are then finally recombined in the computer to form a complex value $S(t)$.

The signal obtained by the receiver coil can be viewed as the combination of the transverse magnetization over the selected slice, expressed mathematically as:

$$S(t) = \iint\limits_{-\infty}^{+\infty} M_{xy} e^{-j\phi(t)} dx\, dy \qquad (A.13)$$

where it is assumed when acquiring the MR signal that the *sampling time* T_s is much less than the relaxation time T_2^* such that the exponential T_2^* decay term, forming

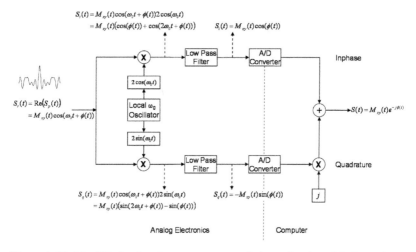

Figure A.17: The MR signal undergoes a process called complex demodulation, using a combination of analogue electronics and computer hardware. The basic steps are separating the raw MR signal, removing the ω_0 carrier frequency, filtering and digitally sampling the signal and finally recombining the two signals to form a complex signal.

part of the transverse magnetization solution given in (A.11), can be ignored. The phase difference is given by:

$$\phi(t) = \int\limits_0^T \omega \, dt = \int\limits_0^T \gamma \left(x G_x + y G_y \right) dt \qquad (A.14)$$

where the time 0 is at the middle of the initial 90 $^\circ$ pulse and T is at the end of the data acquisition. By substituting (A.14) into (A.13) the signal becomes:

$$S\left(t \right) = \iint\limits_{-\infty}^{+\infty} M_{xy} e^{-j\gamma \int\limits_0^T (x G_x + y G_y) dt} \, dx dy \qquad (A.15)$$

By defining the two terms inside the exponential of (A.15) as:

$$k_x = \frac{\gamma}{2\pi} \int_0^T G_x dt \qquad\qquad k_y = \frac{\gamma}{2\pi} \int_0^T G_y dt \qquad (\text{A.16})$$

the signal equation becomes:

$$S(t) = M(k_x, k_y) = \int\int_{-\infty}^{+\infty} M(x,y)\, e^{-2j\pi(xk_x + yk_y)} dx dy \qquad (\text{A.17})$$

The signal received as a function of time therefore defines a trajectory through k space depending upon the relations defined in (A.16). As an example, consider the pulse sequence illustrated in Figure A.15 where the magnetic field gradients G_x and G_y are constant when they are turned on. Since the phase encode gradient G_y is turned off before the data is acquired then the k_y integral will be defined from the time when the phase encode gradient was turned on to the time when it was turned off and will be a constant term, defining a *row* in k space. The frequency encode gradient is turned on while the data is being acquired and hence as the acquisition time proceeds, so too does the position in k space, moving along the row. Figure A.18 illustrates the acquisition of a row of data in k space where, for simplicity the ramp up and ramp down of the read gradient have been ignored. The initial negative dephasing gradient lobe (or rewinder) before the data acquisition begins, is implemented to create a negative k_x trajectory, moving the k_x position from the center (i.e. $k_x = 0$) out to the left k_x limit. As the data is acquired the G_x gradient is switched to a positive value, defining a positive trajectory in k_x. Hence as the data acquisition proceeds in time, the k space trajectory moves from the left most limit to the right most limit and the values of transverse magnetization, sampled as a function of time, undergo the complex demodulation process and are then input directly into the k space matrix. It should be noted that this example illustrates one way to fill k space, but there are a number of other techniques available. Although it is outside the scope of this discussion to cover different k space trajectories in detail, as an example of this point, sinusoidally oscillating G_x and G_y fields can be implemented, which determine a spiral trajectory to cover k space rather than the aforementioned row by row coverage.

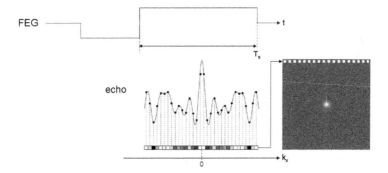

Figure A.18: An example of the acquisition of a line of k space data. The constant read gradient defines a linear trajectory along k_x as the signal is sampled in time. The sampled signal represents the two signals from the complex demodulation process which are entered into k space as complex numbers.

In order to completely fill the k space matrix, the data acquisition cycle needs to be repeated for every row in the k space matrix (i.e. to resolve N locations in the phase encode direction, N different phase encode gradient steps are required), where for each row a different magnitude of the phase encode gradient is applied (Figure A.15). The signal in each row of k space contains the summation of all of the transverse magnetization signal frequencies from the voxels in the selected slice. Returning now to (A.17), it is shown that the MR signal received as a function of time is also a function of k space. Another important result is that (A.17) takes the form of the 2D *Fourier transform* of the transverse magnetization $M(x, y)$. The basic principal behind the Fourier transform is that a complex signal can be decomposed into constituent simple sine and cosine waves of varying amplitude, frequency and phase. This is essentially what is desired, i.e. to take the k space data and separate out the different signal amplitudes from the voxels within the selected slice based on their frequency and phase. Here it is the transverse magnetization as function of spatial location $M(x, y)$ which is desired and this may be obtained using the inverse Fourier transform:

$$M(x, y) = \iint\limits_{-\infty}^{+\infty} M(k_x, k_y) \, e^{2j\pi(xk_x + yk_y)} \, dk_x dk_y \qquad (A.18)$$

However, since the data stored in k space is discrete, the discrete form of the Fourier transform is used:

$$M\left(x,y\right) = \frac{1}{LM} \sum_{k_x=0}^{L-1} \sum_{k_y=0}^{M-1} M\left(k_x, k_y\right) e^{2j\pi\left(\frac{xk_x}{L} + \frac{yk_y}{M}\right)} \tag{A.19}$$

where L and M are the number of entries in the k_x and k_y direction respectively. A given line of data in the k_x direction contains the transverse magnetization signal from the entire selected slice, where each component of the signal has a unique frequency and phase associated with it. It may be intuitive to understand how the fourier transform can be applied in each k_x line to separate out the frequency components, but the separation of phases in the k_y direction is perhaps less intuitive. Considering a given vertical k_y line (Figure A.19), the complex numbers will all have different phases brought about by the application of the different phase encode gradients. The variations in phase along a given k_y line form oscillations in a similar manner to the signal along a line in the k_x direction. As frequency is no more than the rate of change of phase, the application of the Fourier transform to the oscillatory phase components can be thought of in much the same way as that of the frequency data in the k_x direction.

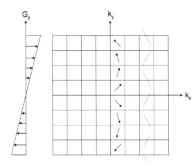

Figure A.19: The MR signal entered into a given k_x line in k space includes the signal from the whole selected slice in both the frequency and phase encode directions. In order to separate the signal in terms of frequency and phase to generate the final MR image phase encode steps of differing phase encode gradient magnitudes are required. Along a given line in k_y, the oscillations in phase form a complex signal (similar to the MR signal read along k_x) allowing the Fourier transform to separate out the different phases.

The result of the inverse Fourier transform of the k space data is an array of

complex numbers containing the summation of all the spatial waves with their varying frequency and phase. Generally the final resulting MR image is the *magnitude
image*, where for a given (x, y) pixel location, the intensity $I(x, y)$ is calculated as:

$$I(x, y) = \sqrt{Re(M(x, y))^2 + Im(M(x, y))^2}$$
(A.20)

Image quality in MRI is a complicated issue, as there are a number of parameters
which may be altered to improve a certain aspect of the image quality, but which
usually involve a tradeoff with another aspect of the image quality. Noise, in the form
of electromagnetic radiation in the body due to movement of charged particles, or
small anomalies in the measurement electronics, is always present in an MR image
and one important criterion for image quality is the *signal to noise ratio (SNR)*.
Image resolution is also an important criterion, which is dependent on the *field of
view (FOV)* of the image and the k space matrix size. While it is desirable to
have high resolution images, generally increasing resolution decreases the SNR. The
reason behind this is that the increase in resolution is accomplished by acquiring
data further out in k space at the higher spatial frequencies, which with reference
to the relations defined in (A.16) can be achieved by either altering the gradient
fields or the length of the data acquisition period. In either case, the increased
resolution is a result of obtaining MR signal from smaller voxels across the FOV,
which means that their will we less signal per voxel, although the noise per voxel
remains relatively constant, hence the SNR decreases. To emphasize via an example,
consider increasing the gradient field strength across the sample, which means that
the protons would be precessing at a larger range of frequencies across the sample.
This would also mean that the *receive bandwidth (rBW)* (the range of frequencies
received by the coil) increases and therefore the MR signal must be sampled more
frequently (in keeping with the *Nyquist criterion* [143]), therefore creating a k space
trajectory that covers a larger region, hence the size of the k space filled with data
also increases. Since the k space matrix size has increased for the same FOV then the
resolution will increase. The decrease in SNR comes from the fact that increasing the
gradient field strength (and hence the range of precessional frequencies of protons
across the sample) means that at any given location across the FOV there will be
fewer protons precessing at a particular frequency and therefore less MR signal at
any given frequency. Since the noise power spectrum is relatively constant over the
range of frequencies, the decrease in signal for the same level of noise means that
overall the SNR decreases.

For the present study, where a 1.5T MRI scanner was used, the attainable resolution, while still producing images with an acceptable SNR was between approximately 0.5-0.7mm. The SNR increases by reducing the FOV or increasing the slice thickness. Another important parameter in terms of image quality is the *number of excitations (NEX)*, meaning the number of images averaged to obtain a final resultant image. Since the noise in an image is random, but the MR signal is not, then averaging the images will tend to cancel out the noise while enhancing the MR signal. In practice the NEX may be between 2 - 4, where increasing the number can produce a better quality image, but poses a tradeoff in terms of imaging time.

A.6   Angiography

The two specialist scans used for the present study are angiographic techniques, meaning that they are designed to image blood vessels, in particular the cerebral arteries. The first type of scan to be explained is known as a *Time of Flight (TOF)* scan, which relies on flow dependent changes in longitudinal magnetization. The second type of scan is known as a *Phase Contrast (PC)* scan, which relies on flow dependent changes in transverse magnetization.

A.6.1   Time Of Flight

The time of flight principle (Figure A.20) functions by using a short repetition time, causing a saturation, such that the MR signal from the tissue of the selected slice is suppressed in a similar manner to that illustrated in Figure A.8. While the stationary tissue experiencing this saturation will have experienced multiple RF pulses, moving tissue such as blood flowing in an artery will not experience as many RF pulses, hence the signal is not suppressed. The amount of fresh, unsaturated blood entering the slice depends upon the velocity of the blood, the thickness of the slice being imaged and the repetition time used, where a faster blood velocity, longer repetition time or a thinner selected slice will all mean that a given *bolus* of blood will completely leave the selected slice between repetition times and fresh blood will enter, giving a stronger signal. The effect of the flowing blood can essentially be thought of as shortening the T_1 characteristics of the tissue, meaning that for a given repetition time a much greater longitudinal magnetization recovery occurs than for stationary tissue.

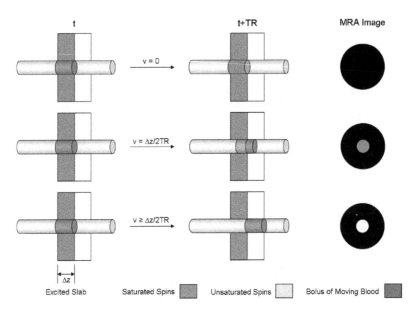

Figure A.20: The time of flight principle. Stationary tissue or blood will experience multiple RF pulses and will become saturated, giving little or no MR signal. Moving blood will not experience same number of RF pulses and will give a stronger signa.l

The time of flight technique will in principal produce an MR signal from both arterial and venous flowing blood. It is usually desired however, to image either one or the other. Since blood in the arteries and veins tends to flow in opposite directions, one method which can used to suppress the MR signal from the unwanted blood vessels is to apply a saturation band immediately before the selected slice (Figure A.21) on the side of the entering blood which is to be suppressed. The saturation band uses RF pulses with a short repetition time, similar to the selected slice and in the case of the present study where it is the arteries which are of interest, the saturation band is applied immediately before the venous blood enters the selected slice.

The discussion to this point has been concerned with selecting a plane that is essentially 2D using the slice encode gradient and generating an image within this slice using the frequency and phase encode gradients. Another method, which is

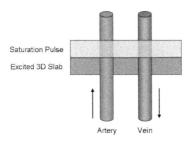

Saturation Pulse

Excited 3D Slab

Artery   Vein

Figure A.21: In order to suppress the signal from venous blood a region above the selected slab has an RF pulse applied to it to saturate the magnetization in the venous blood. As the venous blood enters the excited slab it will hence give little signal. Arterial blood flowing in the opposite direction will not experience the saturation band (before entering the selected slab that is) and will still give a strong signal.

used in a variety of scans, including time of flight of the cerebral arteries, is *3D imaging*. In this case the RF pulse has a larger transmit bandwidth so that a thick slab will be excited, rather than a thin slice. The thick slab is then subdivided into thin slices by using the G_z slice encode gradient as a second phase encode gradient (Figure A.22). 3D imaging has a number of advantages over 2D imaging, beginning with an enhanced signal to noise ratio since much more tissue is excited and therefore able to produce an MR signal. In addition, the time to generate a single image is reduced. The reason for this comes from understanding that although the RF pulse used to excite the protons appears as a *boxcar function* in the frequency domain (i.e. a constant amplitude over the range of frequencies emitted), the signal in the time domain (which is how the signal must be generated physically) is a *sinc function* (i.e. the Fourier transform of the boxcar function). It is a well known result that the wider the boxcar function in the frequency domain, the narrower the sinc function in the time domain and vice versa. Since exciting a thin 2D slice requires a narrow bandwidth RF pulse, the corresponding sinc function is wider in the time domain meaning that it takes longer to produce than a sinc pulse representing a wider bandwidth RF pulse. Finally, the 3D imaging technique, though it seems counterintuitive, allows for much thinner slices to be obtained compared to 2D imaging. This feature arises from the high gradients and low bandwidths required to excite a thin 2D slice, testing the limitations of the hardware. The use of the slice encode gradient as a second phase encode gradient, bypasses these limitations as positions along the z axis can be phase encoded and once separated, provide higher resolution in the slice encode direction. It can also be observed in Figure

A.22 that there is no spin echo used. Instead the echo in the signal is generated by the rewinder (or negative lobe) of the frequency encode gradient, creating a *gradient echo*. In addition to altering the position in k space the rewinder causes the spins to rephase in a similar manner to the echo (in fact rephasing the spins and altering the position in k space are one and the same). The gradient echo has the advantage over a spin echo technique that the echo time can be much shorter, allowing for faster scan times, but has the disadvantage that it does not eliminate the T_2' effects.

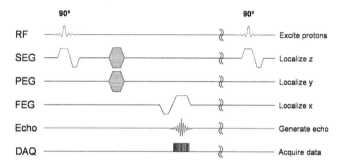

Figure A.22: Pulse and gradient sequence for 3D time of flight imaging uses the wide bandwidth RF pulse to excite a thick slab. In this case the k space becomes three dimensional with k_z being the third dimension. The pulse sequence uses the slice select gradient as a phase encode gradient to control the trajectory through k_z. In addition, the read gradient is used to generate the echo in the MR signal.

The utilization of a second phase encode gradient hence means that the k space becomes a 3D space as opposed to a 2D space for 2D imaging. In order to separate out the magnetization in terms of its 3D spatial location, to generate the final set of images, the 3D Fourier transform is used:

$$M\left(x, y, z\right) = \frac{1}{LMN} \sum_{k_x=0}^{L-1} \sum_{k_y=0}^{M-1} \sum_{k_z=0}^{N-1} M\left(k_x, k_y\, k_z\right) e^{2j\pi \left(\frac{xk_x}{L} + \frac{yk_y}{M} + \frac{zk_z}{N}\right)} \tag{A.21}$$

where N represents the number of phase encode steps and hence the number of slices in the z direction, and k_z is defined in a similar manner to k_x and k_y in (A.16) as:

$$k_z = \frac{\gamma}{2\pi} \int_0^T G_z dt \qquad (A.22)$$

The final set of images from a 3D time of flight scan forms a 3D dataset, which may be used to generate a model of the surface topology of the circle of Willis. A single time of flight image is illustrated in Figure A.23(a), showing the enhanced signal from the blood flowing into the slice. One common technique used to generate an image of the cerebral arteries the *Maximum Intensity Projection* algorithm. Essentially this technique can be thought of as tracing rays through the 3D TOF dataset and interpreting the maximum voxel intensity value encountered by the ray as the corresponding pixel intensity in the maximum intensity projection image, an example of which is shown in Figure A.23(b). In practice more than one 3D slab is usually used, where typically 3 to 4 slabs are used to cover the cerebral vasculature around the region of the circle of Willis.

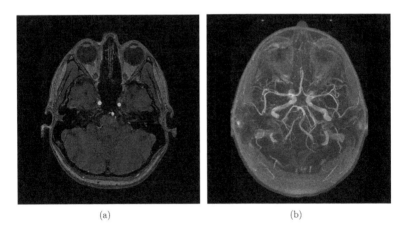

(a)   (b)

Figure A.23: (a) A time of flight image for a transverse slice, illustrating the enhanced signal from the flowing blood and (b) a maximum intensity projection illustrating the arteries comprising the circle of Willis.

A.6.2   Phase Contrast

The phase contrast principal is comprised of creating a phase change in the spins within a given slice using a bipolar gradient (Figure A.24) with the frequency encode (or read) gradient. The total phase change at the time the data is acquired is zero for stationary tissue, but non zero for moving tissue, such as flowing blood. The amount of phase change is directly proportional to the velocity of the blood, hence phase contrast is an inherently quantitative imaging technique, as opposed to the time of flight technique.

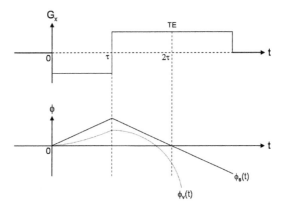

Figure A.24: Phase accumulation for both stationary and constant velocity spins during a bipolar gradient. At the center of the echo the phase accumulated by stationary spins is zero, but for spins moving with constant velocity the phase accumulated in non-zero and is proportional to the velocity of the flow.

To visualize how the phase contrast principle works, consider the position of a bolus of blood $x(t)$, moving in the x direction as a function of time, defined as:

$$x(t) = x_0 + v_x t \qquad\qquad (A.23)$$

where v_x is the constant velocity in the x direction. The phase accumulated by the spins in the selected slice during the negative lobe of the bipolar gradient $\phi_-(t)$ is given as:

$$\phi_-(t) = -\int_0^\tau w(t)dt = -\int_0^\tau \gamma G_x(t)x(t)dt \tag{A.24}$$

where τ is the time at the end of the application of the negative lobe. Neglecting for simplicity the ramp up and ramp down of the bipolar gradient, the accumulated phase is:

$$\phi_-(t) = \gamma G_x x_0 \tau + \frac{1}{2}\gamma G_x v_x(\tau^2) \tag{A.25}$$

Similarly, for the positive lobe, the accumulated phase $\phi_+(t)$ is:

$$\phi_+(t) = -\gamma G_x x_0(t - \tau) - \frac{1}{2}\gamma G_x v_x(t^2 - \tau^2) \tag{A.26}$$

The total accumulated phase is hence:

$$\phi(t) = -\gamma G_x x_0(t - 2\tau) - \frac{1}{2}\gamma G_x v_x(t^2 - 2\tau^2) \tag{A.27}$$

where the first term in (A.27) represents the phase accumulated by stationary tissue and the second term represents the phase accumulated by moving tissue. At the time of the echo, where t is equal to 2τ, the phase accumulated by the stationary tissue is zero and the phase accumulated by the moving tissue is given as:

$$\phi(T_E) = -\frac{1}{2}\gamma G_x v_x \tau^2 \tag{A.28}$$

The pulse and gradient sequence used for a typical 2D phase contrast scan is illustrated in Figure A.25. It can be observed that in practice, two bipolar gradients are used (shown by the solid line and the shaded area) with the sequence of the positive and negative lobes reversed, generating two images for a given slice. The reason for this is that due to magnetic field inhomogeneities (i.e. T_2' effects) an additional phase is accumulated, which is not flow dependent and will result in an artefact in the image. These effects are independent of the sequence of the bipolar gradients however and the T_2' effects may be cancelled by subtracting the two images.

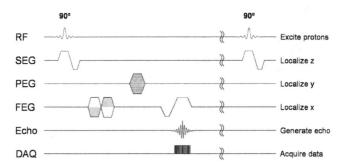

Figure A.25: Pulse and gradient sequence for a phase contrast scan which uses data from two applications of the bipolar gradient (applied in two separate acquisitions), to eliminate phase accumulation due to magnetic field inhomogeneities.

The explanation thus far has considered flow only in the x direction, and it should be apparent that general flow within the cerebral arteries will have velocity components in all three directions. It is therefore convenient to remove the major restriction imposed on the slice, phase and frequency encode gradients thus far; that they coincide with the laboratory frame of reference. Instead the imaging plane (determined by the slice encode gradient) can be created normal to the longitudinal axis of the blood vessel under investigation, using various combinations of the G_x, G_y and G_z gradients. In practice a TOF scan would generally be performed and the maximum intensity projection data used to determine optimal positioning of the imaging plane. Figure A.26 illustrates an example of the specification of an oblique plane, where the azimuthal angle Φ and the polar angle Ψ are defined as:

$$\Psi = \tan^{-1}\left(\frac{\sqrt{G_x^2 + G_y^2}}{G_z^2}\right) \qquad\qquad \Phi = \tan^{-1}\left(\frac{G_y}{G_x}\right) \qquad (A.29)$$

An oblique imaging plane will signify that the phase and frequency encode directions do not coincide with the laboratory frame, although it is still required that they be orthogonal to the slice encode direction. The result of the inverse Fourier

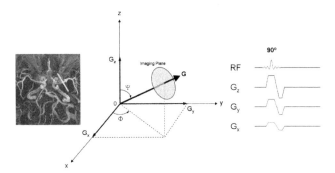

Figure A.26: Variations of the three magnetic field gradients may be combined to create slice select planes at any orientation, relative to the laboratory frame. Typically the TOF data will be used to determine the positioning of the plane. Also shown is a schematic representation of the part of the pulse and gradient sequence used to generate the imaging plane.

transform of the PC data is an array of complex numbers, as with all other MR images. Unlike other imaging techniques however, where it is the magnitude image which is used, the PC scan uses the *phase image*, where for a given pixel location (x, y) the intensity I is calculated as:

$$I(x,y) = \tan^{-1}\left(\frac{Re(M(x,y))}{Im(M(x,y))}\right) \tag{A.30}$$

With most MR images, using the magnitude image, the actual value of the pixel intensity is arbitrary and it is only the relative differences in pixel intensity (representing the differences in tissue contrast) which are of concern. With the phase contrast image however, because the phase image utilizes the ratio of the real and imaginary parts of the complex data produced from the Fourier transform the phase image involves pixel intensities which are in direct proportion to the velocity of the blood and the actual velocity can be calculated by making use of the relation in (A.28). Figure A.27(a) illustrates a sample result of a phase contrast scan of blood vessels, showing the magnitude image. The single resulting phase image shown in Figure A.27(b) is obtained by subtracting the two phase images acquired with the different application of the bipolar gradients, eliminating the effects of magnetic field inhomogeneities. It can be observed that some vessels appear bright where others appear dark, while the surrounding stationary tissue is grey, in the middle of the scale. The reason for this is that the direction of flow is encoded in the phase contrast image and the intensity range includes both positive and negative values of velocity. The bright and dark pixel intensities represent flow in opposite directions and the stationary tissue (having zero velocity) is in the middle of the intensity range and hence appears grey.

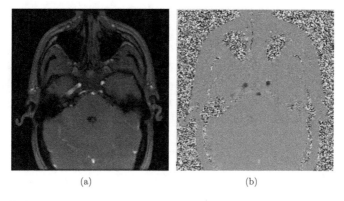

(a)　　　　　　　　　　　　　　　(b)

Figure A.27: An example of a phase contrast scan of some blood vessels showing (a) A magnitude and (b) phase image.

An important point worthy of mention is that technique can only be used for changes in phase of the moving protons up to 180 °s, at which point *aliasing* occurs. In such a case, a positive phase change greater than 180 °s from flow in a given direction can not be distinguished from a negative phase change less than 180 °s from flow in the opposite direction. There is hence an upper limit of the velocity which can be measured in a PC scan, known as the *encoding velocity* v_{enc}, and given as:

$$v_{enc} = \frac{\pi}{\gamma G \tau^2} \tag{A.31}$$

The encoding velocity can be chosen for a particular scan, and so in practice this is not a problem, but does require some knowledge of the expected velocity range in a given vessel. Ideally the encoding velocity should be set as close as possible to the maximum velocity within a blood vessel. If it is set too low, then aliasing will occur, but if it is set too high, so that any aliasing is eliminated, the accumulated phase will be small and hence the intensity in the final phase image will be too low for a reliable calculation.

Considering a single blood vessel, the volume flowrate of blood Q, may be calculated as:

$$Q = \int_{ROI} v dA \qquad (A.32)$$

where v is the velocity of the blood over the region of interest ROI, specified as the cross sectional area of the blood vessel. Using the discrete PC data the flowrate may be approximated by selecting a region of interest (the pixels which comprise the cross section of the blood vessel) and calculating the flowrate as:

$$Q = \sum^{ROI} v_{pix} A_{pix} \qquad (A.33)$$

where v_{pix} is the velocity in a given pixel, A_{pix} is the area of the pixel and N_{pix} is the number of pixels in the region of interest, comprising the blood vessel. For small angles of deviation between the imaging plane and the cross sectional axis of the blood vessel, the calculation of flowrate will not be affected significantly. It follows that if the imaging plane is at an angle θ to the cross section of the blood vessel, then the velocity profile v in the phase image will appear reduced to $v \cos \theta$. The cross sectional area of the blood vessel A, on the other hand will appear increased by the amount $A/\cos \theta$ (i.e. the cross section of the blood vessel will appear as an ellipse in the phase image). When the overall flowrate measurement is made the decreased velocity profile and increase area cancel, yielding an accurate result

One important technique used in conjunction with phase contrast scans used for the present study is *cardiac gating*, where the combination is known as *CINE phase contrast*. With this technique an *electrocardiogram* is taken of the individual undergoing the scan, such that when the MR data is acquired, it is known at what point in the individuals cardiac cycle the data represents. While the actual scan time required to perform a phase contrast scan is on the order of minutes (much longer than the cardiac cycle), if it is assumed that the individuals heart rate remains constant throughout the scan (such that the blood flow is not altered by any factors other than the cardiac cycle), then the MR data which is acquired continuously over a number of minutes can be processed, following the scan and the position of each line of data can be positioned within the cardiac cycle. The result for the phase contrast image is a calculation of the cerebral flowrates over the cardiac cycle.

Appendix B

The Marching Cubes Algorithm

The purpose of this appendix is to provide an outline as to how a 3D MR TOF dataset of the form $M(x, y, z)$ (where the entries in the array correspond to signal intensity from the tissues) may be used to create a 3D model of the surface topology of the circle of Willis.

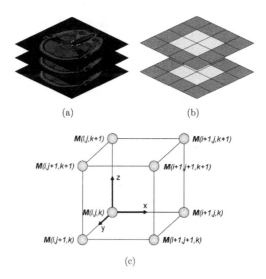

(a)　　　　　　　　(b)

(c)

Figure B.1: The basis for marching cubes illustrated with (a) a stack of MRI images from a 3D TOF scan forming the 3D array M, (b) two consecutive slices in the array with the 4 voxels used per slice highlighted, (c) the eight array values used to form the vertices of the cube with their corresponding assigned values from the 3D array M.

The basis for the algorithm used for the present study is the *Marching Cubes* algorithm developed by Lorensen and Cline [92]. The basic idea behind the method is the definition of a cube where the eight vertices forming the cube are assigned the values of corresponding entries in the 3D array. To illustrate this idea further, Figure B.1(a) shows a stack of MRI time of flight images, forming the 3D array. By then 'zooming' in on the dataset, the eight vertices from two sequential slices (Figure B.1(b)) are used to form the cube (Figure B.1(c)), which may be *marched* through the 3D array, being reassigned the appropriate values to its eight vertices at every position.

In order to generate a surface topology, the next step involves the specification of an *isosurface* intensity threshold, corresponding to the MR signal intensity from the lumen of the circle of Willis. As the cube is marched through the 3D array, the values assigned to its vertices are compared to the isosurface value. If all of the vertices are either above or below the isosurface value then the cube must lie either entirely inside or outside of the lumen respectively. If however some of the vertices are above the isosurface value and some are below, then the boundary of the lumen (the surface of the model) must pass through that cube and will therefore intersect the edges of the cube. This surface may be defined by a *surface triangulation*, which is a collection of connected planar triangles, approximating the smooth arterial wall. A surface triangulation can be described by the combination of a *Faces* and *Vertices* array:

$$
Faces = \begin{vmatrix} V_1 & V_2 & V_3 \\ V_1 & V_3 & V_4 \\ V_2 & V_3 & V_5 \\ & \cdot & \\ & \cdot & \\ & \cdot & \end{vmatrix} \qquad Vertices = \begin{vmatrix} x_1 & y_1 & z_1 \\ x_2 & y_2 & z_2 \\ x_3 & y_3 & z_3 \\ x_4 & y_4 & z_4 \\ x_5 & y_5 & z_5 \\ & \cdot & \\ & \cdot & \end{vmatrix} \qquad \text{(B.1)}
$$

where each row of *Vertices* corresponds to a single vertex V_i and stores its x, y, z coordinates in that row. Each row of *Faces* corresponds to a single face F_i and stores the corresponding row entries in *Vertices* of the 3 vertices which define it. This approach to defining a triangulation has the advantage that a given vertex can be shared by multiple faces, reducing the amount of memory required to store the triangulation. The task then remains to define the surface passing through a cube.

Given that any vertex is either above or below the isosurface, then there are 2^8 or 256 possible combinations for the surface intersecting the cube (i.e. 256 ways in which it may be triangulated). If rotational symmetries are taken into account and furthermore it is understood that the same surface triangulation of a given cube would result if the values of its vertices were reversed (i.e. all vertices above the isosurface were switched to below and vice versa), known as *complementary symmetries*, then the number of combinations reduces to 15 basic families (Figure B.2).

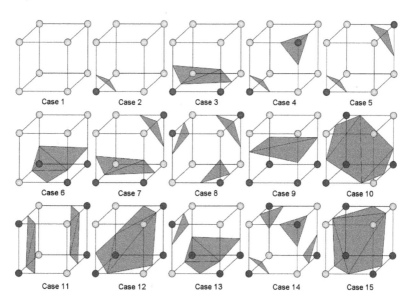

Figure B.2: The 15 families used in the original marching cubes algorithm [92]. These families account for every possible way that a cube can be triangulated, if rotational and complementary symmetries are considered for each case. The light blue and dark blue vertices represent vertex values that lie on opposite sides of the isosurface values (i.e. dark blue is above and light blue below, or vice versa).

The original marching cube algorithm made use of these families by assigning an 8bit index to each cube where each bit in the index corresponded to one of the vertices of the cube and could be 1 if the vertex was above the isosurface and 0 if below. This index would then act as a pointer into an *edge table*, which would store the definitions of the triangles for each particular case. The exact intersection point along of the surface with an edge of the cube can be determined through linear interpolation using the two vertices defining the edge and the isosurface value, meaning that if the isosurface value is numerically close to the value at one of the two vertices then the surface will intersect the edge closer to that vertex and vice versa. As illustrated in the example of Figure B.3(a), the edges of the surface triangulations between adjacent cubes generally connect. In using the edge table, there do arise certain ambiguous cases however, where the surface triangulations assigned to two adjacent cubes do not connect properly, resulting in *gaps* in the surface (Figure B.3(b)). Since these ambiguities only occur with certain families (e.g. cases 4 and 7) [166] then this problem can be remedied by defining *complementary cases* which may be used in place of these ambiguous cases to produce a connected surface, re-triangulating the ambiguous cubes using a *marching tetrahedra* algorithm [150] or asymptotic decider algorithms to resolve the ambiguity [113].

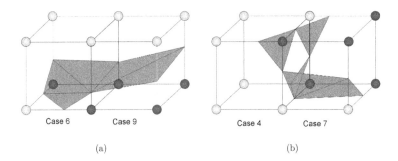

Figure B.3: (a) An example of a general case where the surface triangulation between adjacent cubes produces a continuous surface triangulation and (b) an example of an ambiguous case which can occur with the 15 families. Although these two cases triangulate their own cube appropriately, the triangulations do not connect across the cubes, resulting in a surface triangulation with gaps in it.

The algorithm implemented for the present study is based on the *MATLAB isosurface* function, but has been optimized for the present study to increase the computational speed and the quality of the surface triangulation. While using the idea of defining and triangulating a cube which is marched through the 3D array, this algorithm differs from the original marching cube algorithm in that the edge table is not used to store the definitions of the triangulations, rather each cube is independently triangulated, eliminating the problem of ambiguous cases. To begin the explanation of the algorithm the standard cube needs to be defined. Figure B.4(a) illustrates the convention for the numbering of the vertices and edges, where edges E are numbered from 1 - 12 and vertices V are numbered from 1 - 8.

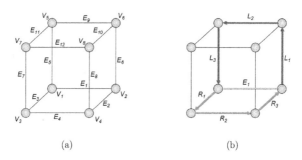

(a)   (b)

Figure B.4: (a) The labelling scheme used for a standard cube, where E designates an edge and V a vertex (b) an example of the left edges L_1, L_2, L_3 and right edges R_1, R_2, R_3 for Edge 1.

In a similar manner to the original marching cubes algorithm, the present method utilizes the definition of an edge table, but differs in that it stores connectivity information between edges of the cube. As illustrated in Table B.1 each row in the edge table for the cube defined in Figure B.4(a) corresponds to one of the 12 edges. The first part of the edge table contains the two vertices V_{n1} and V_{n2} which define it and the next two parts contain the neighbouring *left edges* and *right edges* L_{n1}, L_{n2}, L_{n3} and R_{n1}, R_{n2}, R_{n3} respectively, which complete the definition of the two faces of the cube connected to that edge (Figure B.4(b)). The reason for the definition of the left and right edges will become apparent during the description of the algorithm, but their function is essentially to provide the connectivity information for the surface triangulation.

To aid in the visualization of the algorithm, the forthcoming explanation will be accompanied by way of example of the triangulation of two cube configurations. The first (Figure B.5(a)), is the simplest of all of the possible combinations of a surface intersecting the cube and the second (Figure B.5(b)) is one of the more complicated. During the marching procedure through the 3D array, the algorithm initially performs an *edge loop* over 12 edges defining the cube, calculating the

Table B.1: The edge lookup table used in the present algorithm. Each edge is defined by its two vertices and its left and right edges.

Edge	Vertices	Left Edges	Right Edges
E_n	V_{n1}, V_{n2}	L_{n1}, L_{n2}, L_{n3}	R_{n1}, R_{n2}, R_{n3}
E_1	V_1, V_2	E_6, E_9, E_5	E_3, E_4, E_2
E_2	V_2, V_4	E_8, E_{10}, E_6	E_1, E_3, E_4
E_3	V_1, V_3	E_4, E_2, E_1	E_5, E_{11}, E_7
E_4	V_3, V_4	E_2, E_1, E_3	E_7, E_{12}, E_8
E_5	V_1, V_5	E_{11}, E_7, E_3	E_1, E_6, E_9
E_6	V_2, V_6	E_9, E_5, E_1	E_2, E_8, E_{10}
E_7	V_3, V_7	E_{15}, E_8, E_4	E_3, E_5, E_{11}
E_8	V_4, V_8	E_{10}, E_6, E_2	E_4, E_7, E_{12}
E_9	V_5, V_6	E_{11}, E_{12}, E_{10}	E_5, E_1, E_6
E_{10}	V_6, V_8	E_{12}, E_{11}, E_9	E_6, E_2, E_8
E_{11}	V_5, V_7	E_7, E_6, E_5	E_9, E_{10}, E_{12}
E_{12}	V_7, V_8	E_8, E_4, E_7	E_{11}, E_9, E_{10}

intersection distance along the nth edge d_n as:

$$d_n = \frac{i_{so} - M(V_{n1})}{M(V_{n2}) - M(V_{n1})} \tag{B.2}$$

where i_{so} is the isovalue. d_n has its point of origin at V_{n1} for each edge and based on this calculation, a number of features of the surface's intersection with the cube can be determined. If the values of M assigned at the vertices V_{n1} and V_{n2} defining a given edge are equal, then that the edge can not be intersected by the surface. This fact is reflected by the fact that the denominator of (B.2) would be zero and d_n therefore infinite. If the two values at V_{n1} and V_{n2} differ, then the denominator will be non-zero (either positive or negative). If the isovalue lies in between these two numbers (i.e. $V_{n1} < i_{so} < V_{n2}$ or $V_{n2} < i_{so} < V_{n1}$) then d_n will be a positive fraction in the range of 0 - 1 and it is when d_n evaluates as such that a given edge can be determined to be a *valid* intersected edge. however, if the isovalue lies outside the range of these two numbers (i.e. $V_{n1} \not< i_{so} \not< V_{n2}$ or $V_{n2} \not< i_{so} \not< V_{n1}$) then d_n will evaluate to be either less than zero, or greater than one, but in either case the edge can be determined to be *not* a valid intersected edge. All of the valid intersected edges of the cube have two properties assigned to them, the first is the intersection distance and the second is its list of *next edges* N_1, N_2, N_3. The next edges is used at a later stage in the algorithm and provides a list of edges for a given valid edge to search in order to find other valid edges, so that the points of intersection along valid edges can be connected together. If the sign of the denominator in (B.2) is positive, the next edges are set as the right edges R_1, R_2, R_3 whereas if the sign of the denominator is negative the next edges are assigned to be the left edges L_1, L_2, L_3 of the valid edge. As an example to illustrate this point, consider the case where two vertices are assigned the arbitrary values of 1 and 3, with an isosurface value chosen as 2, then it should be obvious that the edge will be intersected as the isovalue lies between the two vertex values. If $M(V_{n1}) = 1$ and $M(V_{n2}) = 3$ then d_n will evaluate as 0.5 with the denominator being 2 and the right edges will be assigned as the next edges. Alternatively, if $M(V_{n1}) = 3$ and $M(V_{n2}) = 1$ then the denominator of (B.2) will be -2, although d_n will still evaluate as 0.5 since the numerator of (B.2) will also be negative and the left edges will be assigned as the next edges.

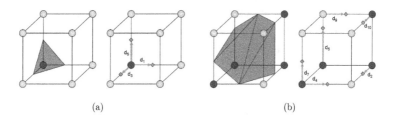

(a)   (b)

Figure B.5: Two example cube configurations illustrating the surface triangulation and the edge intersection distances d_n for each. (a) is the simplest triangulation possible with $N_{valid} = 3$ and (b) is one of the more complex triangulations with $N_{valid} = 6$.

Following the initial edge loop, if there were any valid edges (the total number of which is defined N_{valid}), then a second edge loop is performed to generate the triangles composing the intersected surface. As this loop begins, the edges are examined in order for validity until one is found. Given the two example cases being outlined, the first valid edge encountered would be E_1 for the simple case (Figure B.6(a)) and E_2 for the complicated case (Figure B.6(b)). At this point the edge under examination is designated as the *current edge* and is added to the collection of edges that will form the sides of a polygon, designated as a *polygon string P*, where:

$$P = \{E_{s1}, E_{s2}, ...\} \tag{B.3}$$

where E_{s1} corresponds to the first edge in the polygon string, E_{s2} the second and so on. It is important to note that E_{s1} is the first valid edge found, which will depend upon the particular configuration of the cube. It will *not* necessarily be E_1. The polygon string is essentially a list of all of the edges which will have their intersection points used to form the vertices of a polygon. The current edge is subsequently reassigned as invalid, so that it cannot be used by another polygon. The algorithm continues by sequentially examining the next edges of the current edge N_1, N_2, N_3, checking them for validity. When another valid edge is found it is assigned as the current edge and the former current edge is assigned as invalid so that it cannot be used again to form the vertex of a polygon. The newfound current edge is added to the polygon string and its next edges are sequentially examined for validity until one is found, with the process repeating until the newfound current edge is the initial edge at the start of the polygon string (Figures B.6(a) and B.6(b)). When this occurs the polygon is closed, with the number of edges comprising it defined as N_{edges}. It should be noted that although $N_{valid} = N_{edges}$ for the two example cube triangulations presented, this will not be true in cases where the cube contains multiple disconnected triangles (cases 4, 5, 7, 8, 11, 13 and 14 in Figure B.2).

If at this point $N_{edges} = 3$ (as would be the situation for the simple cube case of Figure B.5(a)), then the polygon string defines a triangle, which can therefore be directly added to the *Faces* and *Vertices* arrays in B.1. The actual spatial coordinates of the intersection points along a given edge are determined by making use of a *coordinate table* (Table B.2) which can be used to relate the intersection distance along an edge d_n (i.e. a value in the range 0 - 1) to its spatial location, based on the current i, j, k position of the cube within M. The algorithm first examines the *Vertices* array to check if any of these coordinates have already been created during the processing of previous cubes and if so records where in the *Vertices* they occurred. If not then these coordinates are added to the end of *Vertices* as a new row entry. The three row positions of the coordinates defining the face are then added to the end of the *Faces* array as a new row entry, completing the definition of the triangle.

If the length of the polygon string P is greater than 3 (as would be the situation for the more complicated cube case of Figure B.5(b)), then it requires tessellating into a collection of triangles in order to define the surface triangulation for the cube.

The tessellation begins by defining three vertices v_1, v_2, v_3 (which will correspond to the intersection points along the edges of P) and initializing them with the d_n

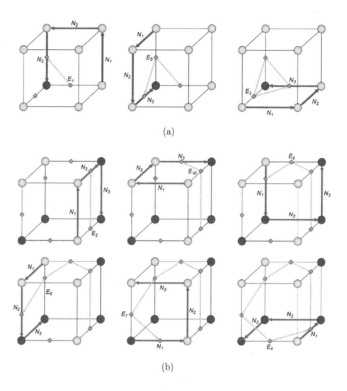

(a)

(b)

Figure B.6: The procedure for adding edges to the polygon string P sequentially examining the next edges of the current edge until another valid edge is found. The process repeats until the edge at the beginning of the polygon string is found. (a) The polygon string is in fact a single triangle with $N_{edges} = 3$ and (b) the polygon string is an arbitrary polygon with $N_{edges} = 6$.

values of the edges stored in P_1, P_2 and P_3. The triangle thus defined with this initialization is added to the $Faces$ and $Vertices$ arrays (using the same procedure outlined previously) and a *polygon string loop* (Figure B.7) is then performed over the remaining edges in the polygon string (i.e. $N_{edges} - 3$ times). For each loop the vertex v_2 is exchanged for v_3, v_3 is reassigned the next entry in the polygon string and the resulting triangle is added to $Faces$ and $Vertices$ until the last edge in the polygon string P_{end} has been assigned to a triangle. It can be observed in Figure B.7 that v_1 is not reassigned during the polygon loop and the 3D spatial coordinate associated with this vertex will therefore be used by every triangle in the

Table B.2: The coordinate lookup table used in the present algorithm to convert the intersection distance along an edge to 3D coordinates based on the i, j, k position of the cube within the 3D array.

Edge	Intersection Point Coordinates		
E_n	v_x	v_y	v_z
E_1	$i + d_1$	j	k
E_2	$i + 1$	$j + d_2$	k
E_3	i	$j + d_3$	k
E_4	$i + d_4$	$j + 1$	k
E_5	i	j	$k + d_5$
E_6	$i + 1$	j	$k + d_6$
E_7	i	$j + 1$	$k + d_7$
E_8	$i + 1$	$j + 1$	$k + d_8$
E_9	$i + d_9$	j	$k + 1$
E_{10}	$i + 1$	$j + d_{10}$	$k + 1$
E_{11}	i	$j + d_{11}$	$k + 1$
E_{12}	$i + d_{12}$	$j + 1$	$k + 1$

tessellation.

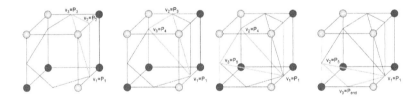

Figure B.7: The procedure for tessellating a polygon string with more than 3 edges. 3 vertices v_1, v_2, v_3 and are initialized with P_1, P_2, P_3 and a triangle is generated with these vertices. The algorithm then loops as many times as there are edges remaining in the polygon string, assigning the value of v_3 to v_2, reassigning v_3 to the next unused edge in P and adding the resulting triangle to the $Faces$ and $Vertices$ arrays.

On completion of this loop, the polygon has then been completely tesselated into triangles. If the number of edges used in by the polygon N_{edges} is equal to N_{valid}, then this means that the cube has been completely triangulated. If however N_{valid} is greater than the amount processed in the polygon, then this means that there will be more than one connected surface passing through the cube (cases 4, 5, 7, 8, 11, 13 and 14 in Figure B.2). The algorithm continues sequentially examining each edge for validity until one is found, at which point it is assigned as the current edge and added into a new polygon string (bearing in mind that the edges which were just used for the previous polygon string have now been made invalid, so that they can't be used more than once). The process of examining the next edges of the current edge until a close polygon is formed and tessellating the polygon string into triangles is repeated until the number of edges which have been used in all of the polygon strings is equal to N_{valid}. At this point the whole cube has been triangulated.

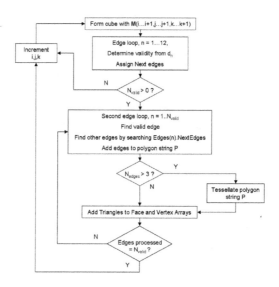

Figure B.8: An outline of the basic steps performed in the marching cubes algorithm used in the present study.

To summarize on the algorithm, Figure B.8 outlines the major steps in the loop. It is important to note that for a given dataset dimensions L, M, N in the x, y, z directions respectively (which corresponds to the k space matrix size), then the numerical values of the 3D coordinates in the vertex array will fall somewhere in the range (i.e. $1 < v_x < I, 1 < v_y < J, 1 < v_z < K$). These coordinates alone are of no meaning, but if the physical spacing between pixels in an image and between slices in an array are known (as is the case with MRI data), then these properties of the scan may be used to subsequently scale the respective x, y, z coordinates of the vertices array so that a scale model of the circle of Willis can be obtained. To end this discussion, Figure B.9 illustrates an example of the posterior afferent portion of a circle of Willis, highlighting the connected surface triangulation produced by the marching cubes algorithm.

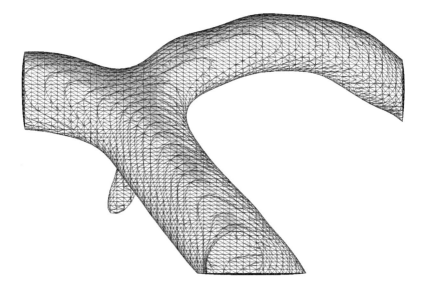

Figure B.9: An example of a surface triangulation of a 3D TOF array, illustrating a portion of the circle of Willis.

Appendix C

Fluid Dynamics

C.1 Introduction

This purpose of this appendix is to provide an outline of the fluid mechanics required to model the cerebral hemodynamics of the circle of Willis. Initially, the mathematical considerations for developing the governing equations will be covered and the equations themselves will be derived from first principles. Subsequently, an explanation of the discretization of the governing equations will be given, to provide some insight as to how these equations are solved numerically.

C.2 Governing Equations

The first major consideration when attempting to model fluid motion is how to treat the fluid substance mathematically. All fluids are composed of individual atoms or molecules and hence one modelling approach, known as the *molecular dynamics* approach, is to treat the fluid system of interest as a collection of molecules and apply Newtons laws of motion directly to each molecule. Considering that in just 1cm^3 of air there are of the order of 10^{19} molecules however, even with the most powerful computers the enormous amount of computation required to track each and every particle in a fluid system means that in most applications the molecular dynamics approach is not yet feasible. Another approach is to instead assume that the fluid is a continuous substance. This assumption is known as the *continuum approximation* and it is presumed that the fluid may be divided up an infinite number of times and still remain a continuous substance. Rather than track the properties of each particle in the system, the continuum approximation yields volume averaged properties of the fluid such as pressure, temperature or density. The continuum approximation is valid in most applications of engineering interest, the criteria being that the physical dimensions of the problem do not approach the same order of magnitude as the molecular dimensions of the fluid. Rarefied gas flows such as the flow around a space vehicle in the upper atmosphere, where the mean free path of the gas molecules can approach dimensions of the vehicle (on the order of metres), illustrate one scenario where the continuum approximation can break down. Furthermore in the study of *microfluidics*, where fluid flow is modelled on an extremely small scale, illustrates another scenario where the continuum approximation breaks down. When dealing with blood flow, the continuum approximation is valid as long as the dimensions of the problem of interest do not approach the scale of the red blood cells, at which point the blood cannot be considered a homogeneous liquid but rather a suspension of particles. As the diameter of a red blood cell is approximately 8μm and the diameters of the arteries of interest for the present study are in the range of $1 - 5$mm, the physical dimensions of the problem are approximately three orders of magnitude higher than the dimensions of the red blood cell, hence the continuum approximation may be applied. A third approach to modelling fluid behavior is a *phenomenological approach*. This approach differs from the previous two in that rather than attempt to represent the fluid behavior with an underlying mathematical theory, empirical data is used to which mathematical functions may be fit. This phenomenological approach is important when developing viscosity models for the complex non-Newtonian nature of blood.

Another major consideration when developing the mathematical models of fluid dynamics is how to represent the fluid system. The choice of system will affect the way the governing equations are derived, and two descriptions exist. The first is to define a region within the fluid and track its 'fate' as it moves with the flow. As the region moves its boundaries may deform but it is a *closed system* as the region always consists of the same fluid particles. This is known as a *Lagrangian description* (Figure C.1). The alternative is to define a region which is fixed in space and allow the fluid to move through it. In this case the region is an *open system* as fluid particles can move into and out of the region. This is known as an *Eulerian description*.

A final consideration involves the definition of the region of fluid and again there are two possibilities. The first is to select an *infinitesimal fluid element* within the flow field (Figure C.1). In this case it is assumed that the element is small enough that the mathematical limiting processes of calculus are valid, but is large enough that the element contains a sufficiently large number of molecules to maintain the continuum approximation. Furthermore it must be assumed that the fluid properties are continuous throughout the spatial domain. A second approach is to select an arbitrary and finite sized volume from within the flow field, known as a *control volume*. When a fluid element is used, the mathematical model takes the form of a system of differential equations. If the control volume approach is used, a system of integral equations results. Generally, the differential form is used when it is desired to have a detailed description of properties throughout the fluid domain, given by the continuous functions which are the solutions to the differential equations. The integral form is used whenever the gross behavior over the control volume boundaries is of interest, instead of the variation of fluid properties within the control volume. It is important to note that whichever approach one uses to develop the governing equations of fluid motion, *all* descriptions of the flow field, whether Eulerian or Lagrangian, differential or integral form, describe the same physical flow field and the equations resulting from each approach can hence be transformed between the various forms through simple mathematical manipulations.

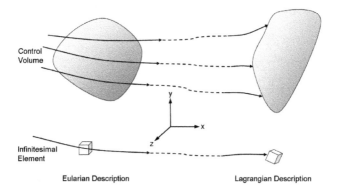

Figure C.1: Eulerian and Lagrangian viewpoints applied to a control volume and an infinitesimal fluid element. The Lagrangian viewpoint involves tracking the respective region of fluid from the left to right with the flow and no mass crosses the system boundary. The system boundary with the Eulerian description is fixed in space and fluid moves into and out of the system.

For the present study the control volume approach is used to develop the gov-

erning equations because it is the integral form of the resulting equations which the
discretization method is based on. The governing equations result from the application of the physical principals of the conservation of mass, momentum and energy
to the control volume. Since the resulting three equations all have a similar form
it is convenient to develop a generic equation for an arbitrary property of the fluid,
using the *Reynolds Transport Theorem*, and subsequently substitute the properties
of mass, momentum and energy to arrive at the final form of the equations. Before
proceeding an important operator in fluid mechanics must be introduced, namely
the *total* or *Stokes* derivative:

$$\frac{D}{Dt} \equiv \frac{\partial}{\partial t} + u\frac{\partial}{\partial x} + v\frac{\partial}{\partial y} + w\frac{\partial}{\partial z} = \frac{\partial}{\partial t} + \mathbf{u} \cdot \nabla \tag{C.1}$$

where $\mathbf{u}$ is the 3D velocity vector containing components u, v, w in the x, y and z
directions respectively. The total derivative represents the total rate of change of a
fluid property and includes two terms. The first term is an *unsteady term*, resulting
from the temporal rate of change of a fluid property. The second term is known as
the *convective term* and arises due to the change in a fluid property from a change
in spatial location within the fluid flow.

C.2.1   Reynolds Transport Theorem

Consider the Eulerian control volume CV located within the flow field of Figure C.2
and bounded by the solid line which represents the *control surface (CS)*. At a certain
time t_0 the fluid system is contained entirely within the control volume and there will
be certain quantities of *extensive* properties such as mass, momentum and energy
contained within it. All of these properties may be generalized by introducing the
arbitrary variable Φ. At some time $t_0 + \Delta t$ later the Lagrangian system will have
moved with the flow and the system boundaries (shown by the dashed line) will no
longer coincide with the control volume. The total rate of change of the property Φ
for the system is given as:

$$\frac{D\Phi}{Dt}\bigg|_{system} = \lim_{\Delta t \to 0} \left(\frac{\Phi_{system}(t_0 + \Delta t) - \Phi_{system}(t_0)}{\Delta t} \right) \tag{C.2}$$

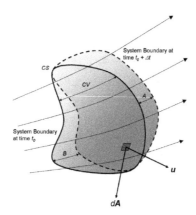

Figure C.2: A flow field moving through a control volume.

where within the limit as $\Delta t \to 0$, the Lagrangian system will coincide with the Eulerian control volume [68]. At the time $t_0 + \Delta t$ the system is composed of the regions $\Phi_A + \Phi_{CV} - \Phi_B$ and hence the total rate of change becomes:

$$\frac{D\Phi}{Dt}\bigg|_{system} = \lim_{\Delta t \to 0} \left(\frac{\Phi_{CV}(t_0 + \Delta t) + \Phi_A(t_0 + \Delta t) - \Phi_B(t_0 + \Delta t) - \Phi_{CV}(t_0)}{\Delta t} \right)$$
(C.3)

which may be separated into:

$$\begin{aligned} \frac{D\Phi}{Dt}\bigg|_{system} = & \lim_{\Delta t \to 0} \left(\frac{\Phi_{CV}(t_0 + \Delta t) - \Phi_{CV}(t_0)}{\Delta t} \right) \\ & + \lim_{\Delta t \to 0} \left(\frac{\Phi_A(t_0 + \Delta t)}{\Delta t} \right) \\ & - \lim_{\Delta t \to 0} \left(\frac{\Phi_B(t_0 + \Delta t)}{\Delta t} \right) \end{aligned}$$
(C.4)

The first term on the right hand side of (C.4) represents the accumulation of the property Φ within the control volume and becomes:

$$\lim_{\Delta t \to 0} \left(\frac{\Phi_{CV}(t_0 + \Delta t) - \Phi_{CV}(t_0)}{\Delta t} \right) = \frac{\partial \Phi_{CV}}{\partial t} = \frac{\partial}{\partial t} \int_{CV} \rho \phi \, dV \qquad \text{(C.5)}$$

which is the time rate of change of the volume integral of the intensive form of the property ϕ and ρ is the fluid density. The two remaining terms on the right hand side of (C.4) represent fluxes of ϕ into and out of the system respectively which becomes:

$$\lim_{\Delta t \to 0} \left(\frac{\Phi_A(t_0 + \Delta t)}{\Delta t} \right) - \lim_{\Delta t \to 0} \left(\frac{\Phi_B(t_0 + \Delta t)}{\Delta t} \right) = \dot{\Phi}_{out} - \dot{\Phi}_{in} = \int_{CS} \rho \phi \, \mathbf{u} \cdot d\mathbf{A} \quad \text{(C.6)}$$

where $d\mathbf{A}$ is the elemental area vector, of magnitude dA and normal to the control surface CS The total rate of change for the extensive property Φ of the system thus becomes:

$$\frac{D\Phi}{Dt}\bigg|_{system} = \frac{\partial}{\partial t} \int_{CV} \rho \phi \, dV + \int_{CS} \rho \phi \, \mathbf{u} \cdot d\mathbf{A} \qquad \text{(C.7)}$$

which is the *Reynolds Transport Equation* [83]. Using (C.7) to apply the principals of conservation of mass, momentum and energy involves the substitution for the intensive property ϕ:

$$\Phi = \left(\begin{array}{c} mass \\ momentum \\ energy \end{array} \right) \quad \phi = \left(\begin{array}{c} 1 \\ \mathbf{u} \\ e \end{array} \right) \qquad \text{(C.8)}$$

where e is the internal energy, which may be related to the fluid temperature T.

The total rate of change of Φ for the three respective cases is thus:

$$\frac{D\Phi}{Dt}\bigg|_{system} = \begin{pmatrix} 0 \\ \sum \mathbf{F}_{external} \\ \dot{Q} - \dot{W} \end{pmatrix} \tag{C.9}$$

where $\sum \mathbf{F}_{external}$ represents the sum of the external forces acting on the control volume and $\dot{Q}$ and $\dot{W}$ represent the heat and work input and output from the fluid respectively. The conservation of mass principal means physically that the amount of mass accumulating within the control volume is equal to the net rate of mass flow into it. The conservation of momentum principal is Newtons second law of motion which states that the sum of the external forces on the control volume is equal to the rate of change of momentum within it. The conservation of energy principal is the first law of thermodynamics which states that the heat input and work output is equal to the rate of change of internal energy.

For a compressible non-Newtonian fluid there are seven variables required to completely describe the flow; the fluid density ρ, the three velocity components u, v and w, the thermodynamic pressure p, the temperature T, and the apparent viscosity η. In order to solve for the seven unknowns, seven equations are required. Five equations can be obtained from the Reynolds transport equation, the remaining two being phenomenological models which relate the fluid viscosity and density to the fluid temperature, pressure and velocity, thereby closing the system. All liquids, including blood, can be treated as incompressible for practical purposes, meaning that the density is decoupled from the fluid pressure. As there is no significant heat transfer within the arteries of the circle of Willis, the phenomenological density and viscosity models need not be functions of temperature. The blood density is hence a constant and the viscosity is decoupled from the fluid temperature. The implication of this is that the energy equation will not affect the flow field in any way, as it is decoupled from the other flow variables. Consequently, only the conservation of mass and momentum equations are required to completely describe the flow (along with the viscosity model) and only the derivation of these equations will be given.

C.2.2   Conservation of Mass and Momentum

Using the Reynolds transport equation, the conservation of mass (continuity) equation becomes:

$$\frac{\partial}{\partial t} \int_{CV} \rho \, dV + \int_{CS} \rho \mathbf{u} \cdot d\mathbf{A} = 0 \tag{C.10}$$

Since blood is being modelled as incompressible however the first term on the right hand side of (C.7) is zero leaving:

$$\int_{CS} \rho \mathbf{u} \cdot d\mathbf{A} = 0 \tag{C.11}$$

For the conservation of momentum equation the Reynolds transport equation gives:

$$\frac{\partial}{\partial t} \int_{CV} \rho \mathbf{u} \, dV + \int_{CS} \rho \mathbf{u} \mathbf{u} \cdot d\mathbf{A} = \sum \mathbf{F}_{external} \tag{C.12}$$

The external forces acting on the fluid can be separated into *body forces*, which act over the whole fluid volume and *surface forces*, which act on the fluid boundaries. Body forces can arise due to gravity, electromagnetic effects or phenomenological lumped parameter effects such as porosity for example. For the present study it is only the last example which will be incorporated as a lumped parameter body force term to couple the effects of *cerebrovascular resistance (CVR)* to the governing equations by defining:

$$\sum \mathbf{F}_{body} = \int_{CV} -\eta \, CVR \, \mathbf{u} \, dV \tag{C.13}$$

which is essentially *Darcy's law* for porous media, assuming laminar flow [33].

Evaluation of the surface forces requires considerably more effort and to begin, it is convenient to consider the infinitesimal fluid element shown in Figure C.3. The

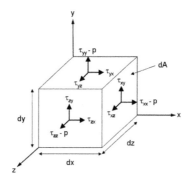

Figure C.3: Stress components acting on an infinitesimal fluid element.

surface forces acting upon the element are caused by the thermodynamic pressure
within the fluid and the effects of viscosity. Thermodynamic pressure will act on
the fluid even if it is not in motion and will always be normal and inward to the
surface of the fluid element. The viscous forces arise due to deformation of the fluid
element and are caused only when the fluid is in motion. The viscous forces create
stresses which can act both normal and tangential to the surface of the element and
the stress components define a second order tensor:

$$
\overline{\overline{\tau}} = \begin{pmatrix} \tau_{xx} & \tau_{xy} & \tau_{xz} \\ \tau_{yx} & \tau_{yy} & \tau_{yz} \\ \tau_{zx} & \tau_{zy} & \tau_{zz} \end{pmatrix}
\tag{C.14}
$$

It now remains to relate the viscous stresses to the velocity components u, v, w
and the viscosity η, using appropriate *constitutive equations* and thereby closing the
system. When a fluid is in motion, every point within the fluid is generally displaced
to a new spatial location with time. It is the relative motion between points within
the fluid which causes deformation. If the velocity vector field is known as a function
of space and time $\mathbf{u}(x, y, z, t)$, then the motion of the fluid is completely described.
There are then *kinematic* expressions for the relative motion between points within
the fluid which can be derived, relating the deformation to the velocity field.

In order to develop the kinematic expressions for the various types of motions

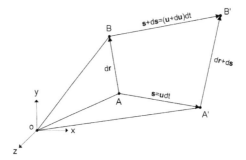

Figure C.4: Relative displacement between two points, A and B within a flow field.

consider the two points A and B existing within a general flow field (Figure C.4) [133]. Due to the presence of the velocity field, point A will be displaced to point A' in a period of time dt by a distance $\mathbf{s} = \mathbf{u}dt$, where $\mathbf{u}$ is the velocity vector at point A. Similarly point B will be displaced to a point B' during the period of time dt, but since B is separated in space from A by a distance $d\mathbf{r}$, the velocity field at that point will be $\mathbf{u} + d\mathbf{u}$ and hence the displaced distance will be $\mathbf{s} + d\mathbf{s} = (\mathbf{u} + d\mathbf{u})\,dt$. Taking a first order Taylor expansion for the velocity vector at point B, the components are:

$$
\begin{aligned}
u \;+\; du &= u \;+\; \frac{\partial u}{\partial x}dx \;+\; \frac{\partial u}{\partial y}dy \;+\; \frac{\partial u}{\partial z}dz \\
v \;+\; dv &= v \;+\; \frac{\partial v}{\partial x}dx \;+\; \frac{\partial v}{\partial y}dy \;+\; \frac{\partial v}{\partial z}dz \\
w \;+\; dw &= w \;+\; \frac{\partial w}{\partial x}dx \;+\; \frac{\partial w}{\partial y}dy \;+\; \frac{\partial w}{\partial z}dz
\end{aligned}
\tag{C.15}
$$

which may be written in a more compact vector form as:

$$
\mathbf{u} + d\mathbf{u} = \mathbf{u} + \nabla\mathbf{u} \cdot d\mathbf{s}
\tag{C.16}
$$

where the ∇ operator applied to the vector $\mathbf{u}$ is known as the *dyadic product* and defines the second order tensor:

$$\nabla \mathbf{u} = \begin{pmatrix} \dfrac{\partial u}{\partial x} & \dfrac{\partial u}{\partial y} & \dfrac{\partial u}{\partial z} \\[2mm] \dfrac{\partial v}{\partial x} & \dfrac{\partial v}{\partial y} & \dfrac{\partial v}{\partial z} \\[2mm] \dfrac{\partial w}{\partial x} & \dfrac{\partial w}{\partial y} & \dfrac{\partial w}{\partial z} \end{pmatrix} \equiv \dfrac{\partial u_i}{\partial x_j} \tag{C.17}$$

The relative motion between points A and B, responsible for the fluid deformation is $(\mathbf{u} + d\mathbf{u}) - \mathbf{u} = d\mathbf{u}$, hence $\nabla \mathbf{u}$ is known as the *rate of deformation* tensor. A useful result in tensor algebra is that every tensor can be uniquely expressed as the sum of a *symmetric* and an *anti-symmetric* tensor and for reasons that will become apparent shortly, it is convenient to separate $\nabla \mathbf{u}$ into the symmetric *rate of strain tensor* $\bar{\epsilon}$ defined as:

$$\bar{\epsilon} = \begin{pmatrix} \dfrac{\partial u}{\partial x} & \dfrac{1}{2}\left(\dfrac{\partial u}{\partial y} + \dfrac{\partial v}{\partial x}\right) & \dfrac{1}{2}\left(\dfrac{\partial u}{\partial z} + \dfrac{\partial w}{\partial x}\right) \\[2mm] \dfrac{1}{2}\left(\dfrac{\partial u}{\partial y} + \dfrac{\partial v}{\partial x}\right) & \dfrac{\partial v}{\partial y} & \dfrac{1}{2}\left(\dfrac{\partial v}{\partial z} + \dfrac{\partial w}{\partial y}\right) \\[2mm] \dfrac{1}{2}\left(\dfrac{\partial u}{\partial z} + \dfrac{\partial w}{\partial x}\right) & \dfrac{1}{2}\left(\dfrac{\partial v}{\partial z} + \dfrac{\partial w}{\partial y}\right) & \dfrac{\partial w}{\partial z} \end{pmatrix} \tag{C.18}$$

or adopting the *Einstein summation* notation:

$$\epsilon_{ij} = \frac{1}{2}\left(\frac{\partial u_i}{\partial x_j} + \frac{\partial u_j}{\partial x_i}\right) \tag{C.19}$$

The anti-symmetric tensor is known as the *rate of rotation tensor* $\bar{\omega}$ and is defined as:

$$\bar{\omega} = \begin{pmatrix} 0 & \dfrac{1}{2}\left(\dfrac{\partial u}{\partial y} - \dfrac{\partial v}{\partial x}\right) & \dfrac{1}{2}\left(\dfrac{\partial u}{\partial z} - \dfrac{\partial w}{\partial x}\right) \\[2mm] -\dfrac{1}{2}\left(\dfrac{\partial u}{\partial y} - \dfrac{\partial v}{\partial x}\right) & 0 & \dfrac{1}{2}\left(\dfrac{\partial v}{\partial z} - \dfrac{\partial w}{\partial y}\right) \\[2mm] -\dfrac{1}{2}\left(\dfrac{\partial u}{\partial z} - \dfrac{\partial w}{\partial x}\right) & -\dfrac{1}{2}\left(\dfrac{\partial v}{\partial z} - \dfrac{\partial w}{\partial y}\right) & 0 \end{pmatrix} \tag{C.20}$$

or in summation notation:

$$\omega_{ij} = \frac{1}{2} \left(\frac{\partial u_i}{\partial x_j} - \frac{\partial u_j}{\partial x_i} \right) \qquad (C.21)$$

In order to understand the reason behind the separation of the rate of deformation tensor in this manner, it is necessary to consider the physics of the fluid motion. Because of the linearity of (C.16) the velocity at a point within the fluid may be separated into four different types of motion, which may be superimposed to obtain the total motion of the fluid. In order to develop the kinematic expressions to relate the velocity field to the viscous stresses, it is useful to consider these different types of motion separately. The different movements are due to the presence of both the velocity of the fluid, and velocity gradients within the fluid.

The four general types of motion are shown in Figures C.5(a) - C.5(d). While the various motions are physically 3D, they are depicted here in 2D for simplicity. It is important to note that the corners of the fluid element represent points within the flow field, hence when the element's size or shape is altered there is a relative motion between points within the fluid, hence a deformation of the fluid. Figure C.5(a) illustrates the first of the four types of fluid motion, namely a *rigid body translation*. This type of motion is unique among the four types of motion in that it arises due to the presence of the velocity field $\mathbf{u}$, not the velocity gradients contained within $\nabla \mathbf{u}$. In this case the fluid element retains both its size and shape. The second type of motion (Figure C.5(b)) is a *rigid body rotation*, where similar to the rigid body translation the fluid element retains both its size and shape and is not deformed, but the motion arises due to the presence of velocity gradients within the fluid. For small rotations the angle θ_1 is given as:

$$\theta_1 \approx \tan \theta_1 = \frac{\overline{BB'}}{\overline{AB}} = \frac{v(B) \Delta t}{\overline{AB}} \qquad (C.22)$$

where $v(B)$ is the velocity in the y direction at point B. Taking the first order Taylor expansion for the velocity at B:

$$v(B) \approx v(A) + \frac{\partial v}{\partial x} \overline{AB} \qquad (C.23)$$

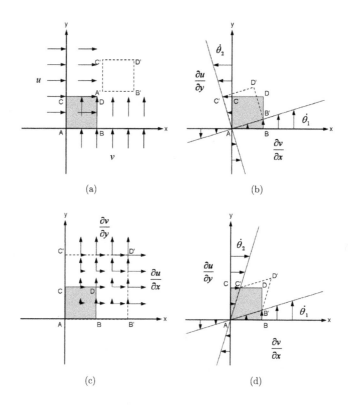

Figure C.5: Types of Motion a Fluid Element May Undergo. (a) Translation, (b) Rotation, (c) Dilation and (d) Distortion

where $v(A)$ is the velocity in the y direction at point A. Noting that $v(A)$ is zero and simplifying (C.23) the angle θ_1 is becomes:

$$\theta_1 = \frac{\partial v}{\partial x} \Delta t \tag{C.24}$$

Dividing by Δt and taking the limit as $\Delta t \to 0$:

$$\lim_{\Delta t \to 0} \frac{\Delta \theta_1}{\Delta t} = \dot{\theta}_1 = \frac{\partial v}{\partial x} \tag{C.25}$$

Using the same reasoning, the time rate of change of the angle θ_2 is:

$$\dot{\theta}_2 = \frac{\partial u}{\partial y} \tag{C.26}$$

On a physical basis the angular velocity of the fluid element about a given axis is defined as the net average rate of anticlockwise spin of the fluid element and for the particular case illustrated in Figure C.5(b), the angular velocity in the xy plane (i.e. about the z axis) ω_{xy} is defined as:

$$\omega_{xy} = \frac{1}{2}\left(\dot{\theta}_1 - \dot{\theta}_2\right) = \frac{1}{2}\left(\frac{\partial v}{\partial x} - \frac{\partial u}{\partial y}\right) \tag{C.27}$$

where it can be observed that (C.27) is a component of the rate of rotation tensor. Extending the motion of the fluid element to 3D, the overall rotation of the fluid element is hence captured by the rate of rotation tensor. The third type of motion (Figure C.5(c)) is known as a *dilation*, where the fluid element retains its shape but its volume may increase or decrease due to the presence of velocity gradients within the fluid. For small deformations, and considering a dilation in the x direction, the rate of dilation will be $\frac{\partial u}{\partial x}$, which is the first component in the rate of strain tensor. Extending the motion of the fluid to 3D, the overall dilation of the fluid element is captured by the *trace* of the rate of strain tensor. The fourth type of motion (Figure C.5(d)) is a *distortion* of the fluid element, where the volume of the element is preserved but its shape is altered by the presence of the velocity gradients within the fluid. Again on physical grounds the rate of strain for the fluid element is defined as the average of the two rates of change of the angle, and using similar reasoning for the rate of change of the angle defined for the rotation in the xy plane, the rate of strain in the xy plane ϵ_{xy} is defined as:

$$\epsilon_{xy} = \frac{1}{2}\left(\dot{\theta}_2 + \dot{\theta}_1\right) = \frac{1}{2}\left(\frac{\partial u}{\partial y} + \frac{\partial v}{\partial x}\right) \tag{C.28}$$

where it can be observed that (C.28) is an off diagonal component of the rate of strain tensor. Extending the motion of the fluid element to 3D, the overall strain is captured by the off diagonal elements of the rate of strain tensor. It should be apparent that only the fluid element dilation and distortion result in a relative displacement between points in the fluid and it is therefore only these two types of motion which will cause the viscous stresses. Furthermore, both of these types of motion are captured within the rate of strain tensor. Since the rate of rotation tensor (associated with the rigid body translation) involves no relative displacement between points in the fluid it will therefore cause no stresses on the fluid element and can subsequently be ignored. It is therefore proposed that the viscous stresses acting on a general fluid element are a function of the rate of strain tensor [133], defined as:

$$\tau_{ij} = \lambda \, \delta_{ij} \epsilon_{ij} + 2\eta \, \epsilon_{ij} \tag{C.29}$$

where δ_{ij} is the kronecker delta, and λ is the *bulk viscosity* of the fluid. For all liquids however, including blood, the density of the fluid is constant and the volume dilation is equal to zero. The first term in (C.29) then becomes zero leaving the constitutive relation between stress and rate of strain as:

$$\tau_{ij} = 2\eta \, \epsilon_{ij} \tag{C.30}$$

For incorporation into the momentum equation the sum of external surface forces is:

$$\sum \mathbf{F}_{surface} = \int_{CS} -p \, I \cdot d\mathbf{A} + \int_{CS} \bar{\tau} \cdot d\mathbf{A} \tag{C.31}$$

where I is the identity matrix. At this point it is convenient to manipulate both the pressure and viscous terms in order to simplify their discretization of the final form of the momentum equation. This simplification is done via *Gauss's divergence theorem*:

$$\int_A \mathbf{q} \cdot d\mathbf{A} = \int_V \nabla \cdot \mathbf{q} \, dV \tag{C.32}$$

where $\mathbf{q}$ may be a vector or tensor quantity and the $\nabla \cdot$ operator indicates the *divergence* of $\mathbf{q}$. Applying the divergence theorem to the viscous term, the stress tensor is transformed to:

$$\int_{CS} \overline{\overline{\tau}} \cdot d\mathbf{A} = \int_{CV} \nabla \cdot \overline{\overline{\tau}} \, dV \tag{C.33}$$

Incorporating the constitutive relation between stress and rate of strain the integral becomes:

$$\int_{CV} \nabla \cdot \overline{\overline{\tau}} \, dV = \int_{CV} 2\eta \nabla \cdot \overline{\overline{\epsilon}} \, dV \tag{C.34}$$

Taking the divergence of the rate of strain tensor and noting the equivalence of mixed derivatives:

$$\begin{aligned}
\nabla \cdot \overline{\overline{\epsilon}} &= \frac{\partial}{\partial x_i} \left(\frac{1}{2} \left(\frac{\partial u_i}{\partial x_j} + \frac{\partial u_j}{\partial x_i} \right) \right) \\
&= \frac{1}{2} \left(\frac{\partial}{\partial x_i} \left(\frac{\partial u_i}{\partial x_j} \right) + \frac{\partial^2 u_j}{\partial x_i^2} \right) \\
&= \frac{1}{2} \left(\frac{\partial}{\partial x_j} \left(\frac{\partial u_i}{\partial x_i} \right) + \frac{\partial^2 u_j}{\partial x_i^2} \right)
\end{aligned} \tag{C.35}$$

Since the $\frac{\partial u_i}{\partial x_i}$ term is the volume dilation $\nabla \cdot \mathbf{u}$, and is zero for an incompressible fluid such as blood, the divergence of the rate of strain tensor becomes:

$$\nabla \cdot \overline{\overline{\epsilon}} = \frac{1}{2} \frac{\partial^2 u_j}{\partial x_i^2} = \frac{1}{2} \nabla^2 \mathbf{u} \tag{C.36}$$

Substituting back into the right hand side of (C.34) and again making use of the divergence theorem:

$$\int_{CV} \eta \nabla^2 \mathbf{u} \, dV = \int_{CV} \eta \nabla \cdot (\nabla \mathbf{u}) \, dV = \int_{CS} \eta \nabla \mathbf{u} \cdot d\mathbf{A} \tag{C.37}$$

Hence, the complete momentum equation thus becomes:

$$\frac{\partial}{\partial t} \int_{CV} \rho \mathbf{u}\, dV + \int_{CS} \rho \mathbf{u}\mathbf{u} \cdot d\mathbf{A} = \int_{CS} \eta \nabla \mathbf{u} \cdot d\mathbf{A} - \int_{CS} p\, I \cdot d\mathbf{A} - \int_{CV} \eta\, CVR\, \mathbf{u}\, dV \quad \text{(C.38)}$$

C.3 Discretization

The governing equations of fluid mechanics form a complex system to which there is generally no analytical solution, apart from a small number of simplified cases. In order to remedy this problem numerical methods are used where the continuous fluid domain is divided into discrete points or regions and the governing equations are then applied to each point or region. The result is a coupled system of algebraic equations which may be solved to provide an approximate solution to the governing equations. There are three basic methods which are used in practice to discretize the governing equations; the *Finite Difference* method, the *Finite Element* method, and the *Finite Volume* method.

The finite difference method is applied to the differential form of the governing equations, whereby the derivatives are replaced by *finite difference* equations at nodal points within the fluid domain, resulting in a system of algebraic equations. The drawback of the finite difference method is that it requires a rectangular *structured grid* with which to apply the finite difference equations. It is still possible to use this method on grids that are not structured as long as it is possible to apply a transformation such that it may be mapped onto a structured grid. As the geometry becomes increasingly more complex, so too does the transformation and in many cases (including the complex geometry of the circle of Willis) it becomes difficult, if not impossible, to calculate them. The remaining two methods are based on the integral form of the governing equations, which have the advantage that they may be used on unstructured grids.

The finite element method is based on the choosing of shape functions, which assume a variation of ϕ within a cell, based on the values stored at its nodal points. Furthermore, by assuming $\overline{\phi}$ to be the approximate solution of ϕ, there will be a *residual* in the governing equations which the finite element method integrates over the fluid domain and sets to zero (in a weighted sense, related to the shape functions) in order to produce a system of algebraic equations.

For the present study it is the last method, namely the finite volume method, which has been chosen and it is the purpose of the remainder of this appendix to explain how it is implemented.

The Reynolds Transport equation, written for an arbitrary variable ϕ (C.7) may be further generalized:

$$\frac{\partial}{\partial t} \int_{CV} \rho \phi \, dV + \int_{CS} \rho \phi \mathbf{u} \cdot d\mathbf{A} = \int_{CS} \Gamma \nabla \phi \cdot d\mathbf{A} + \int_{CV} S \, dV \qquad \text{(C.39)}$$

where Γ is known as the diffusivity and S represents a source of the property Φ. Hence, the four terms in the transport equation are the *unsteady, convective, diffusive,* and *source* terms from left to right respectively. With the continuity and momentum equations developed thus far the diffusivity and source terms are:

$$\Phi = \begin{pmatrix} mass \\ momentum \end{pmatrix} \qquad \phi = \begin{pmatrix} 1 \\ \mathbf{u} \end{pmatrix}$$

$$\Gamma = \begin{pmatrix} 0 \\ \eta \end{pmatrix} \qquad S = \begin{pmatrix} 0 \\ \eta \mathbf{u} \, CVR - \nabla p \end{pmatrix} \qquad \text{(C.40)}$$

where it can be noted that for the continuity equation most of the terms are zero, and for the momentum equation the pressure term has been included with the source term via an application of the divergence theorem. One important issue, worthy of mention at this point, is that of turbulence modelling. Theoretically the conservation of mass and momentum principals encompass all of the physics required to describe fluid flow, including its turbulent nature. Turbulence however, can occur on scales as small as the micron and millisecond spatial and temporal scales respectively, and when the governing equations are discretized over the spatial and temporal domains the spatial grids and the timestep sizes are quite often too large to accurately capture turbulent phenomena. When a particular flow of interest is

known to be turbulent, *turbulence models* may be incorporated as extra equations
to be solved with the continuity and momentum equations. These take the form of
the generalized transport equation (C.39), where the number of separate equations
depends upon the particular turbulence model used. It is known however that blood
flow is laminar throughout the body except in the aortic arch, and occasionally in
regions around stenosed blood vessels. Consequently, the blood flow throughout
the circle of Willis has been assumed to be laminar and no turbulence models have
been implemented for the present study. The usual reason for defining the Reynolds
transport equation is that it can be readily extendible to other governing equations
such as the energy equation, turbulence equations, or species transport equations.

In order to initiate the discretization process it is convenient at this point to
ignore the discretization of the unsteady term in (C.39), focusing on the spatial
discretization of ϕ and then returning to the temporal discretization at a later stage.
The application of the spatially discretized Reynolds transport equation to a single
cell c takes the form:

$$\frac{\partial}{\partial t} \int_{CV} \rho \phi \, dV + \sum_f \rho \phi_f \mathbf{u}_f \cdot \mathbf{A}_f = \sum_f^{N_{facenb}} \Gamma_f (\nabla \phi)_f \cdot \mathbf{A}_f + S_c V_c \qquad (C.41)$$

where the subscripts c and f refer to cell centroid and face centroid values respectively (whose significance will be discussed shortly) and f is summed over the number
of faces N_{facenb} comprising the cell c. The important point to note is that the surface
integrals have been discretized as a summation of the face fluxes across a cell and
the volume integrals have been discretized as a multiplication by the cell volume.
The application of (C.41) to every cell in the domain yields a coupled non-linear
system of algebraic equations. In order to understand how this system is solved it is
most intuitive to initially examine the end matrix form of the equations, and then
delve into how this form is actually generated. Since the face centroid values of a
given variable will ultimately be related to the cell centroid values of neighbouring
cells, the discretized equation applied to a given cell c may be *linearized* to a form
which relates the scalar ϕ_c, stored at the centroid of cell c, to the values ϕ_n stored

at the centroids of its neighbors n:

$$a_c \phi_c = \sum_{n}^{N_{neighbours}} a_n \phi_n + b_c \tag{C.42}$$

where a_c and a_n are coefficients of ϕ_c and ϕ_n respectively and b_c contains 'known' quantities. When the discretized equation is applied to every cell in the domain the resulting system of equations takes the form:

$$A\phi = \mathbf{b} \tag{C.43}$$

where A is a diagonally dominant sparse matrix, and $\mathbf{b}$ is a column vector containing the known quantities and ϕ is the solution vector. Although this resulting system of equations is linear and A could be inverted to provide the solution vector ϕ, the amount of memory required to invert a matrix is usually computationally prohibitive for practical problems. Furthermore, it is an inefficient use of computer memory to store all the entries in a sparse matrix when the majority of the entries, except those centered about the main diagonal, are zero. As a result, iterative methods are used to solve (C.43) and the method used for the present study is a *point-implicit Gauss-Seidel method* [82]. This method involves taking an initial 'guess' at the solution vector ϕ^m and looping over every cell in the computational grid, updating the values of ϕ. The updated value for ϕ_p for the current iterative step $m + 1$ is given by:

$$\phi_c^{m+1} = \phi_c^m + \frac{\alpha}{a_c} \left(\sum_{n=1}^{c-1} a_n \phi_n^{m+1} + \sum_{n=c+1}^{N_{cells}} a_n \phi_n^m + b_c^m \right) \tag{C.44}$$

where N_{cells} is the number of cells in the domain and α is an *under-relaxation factor* for ϕ. While the bracketed term in (C.44) is the standard Gauss-Seidel method for computing the new values of ϕ_c the present study includes under-relaxation. In this case the new value of ϕ is multiplied by an under-relaxation factor α (taking a value between 0 and 1) which is then added to the old value for ϕ. This is to prevent new values of ϕ from altering drastically between iterations. The first part of the bracketed term in (C.44) involves values of ϕ_n from the current iteration step. This is an important feature of the Gauss-Seidel method, that once a new approximation of ϕ_c has been computed, it is used immediately in subsequent calculations. The iteration process may continue until the difference between ϕ^{m+1} and ϕ^m reaches a specified convergence tolerance. If the momentum equation is considered, then ϕ contains the three velocity vector components u, v, and w in the x, y and z directions respectively. Using what is known as a *segregated-implicit* solver, the Gauss-Seidel method is applied to each velocity component separately, resulting in three separate matrix equations to solve. The segregated term means that these flow variables are each solved separately, and the implicit term signifies that each flow variable is related to the value in its neighboring cells.

The transition between the initial application of the discrete transport equation (C.41) and the linearized form in (C.42) requires a substantial amount of work, and it remains to pay some attention as to how the calculation of the a and b terms for every cell is achieved. Before delving in to the complexities however, it is important to understand how the flow field data is stored in the computational grid.

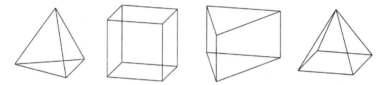

Figure C.6: Tetrahedral, hexahedral, wedge and pyramid cell types used in unstructured grids.

One of the attractive features of the FVM is its applicability to unstructured grids, which are heavily employed in CFD problems. Generally there are four basic types of 3D cells used; *tetrahedral, hexahedral, wedge,* and *pyramid* cells (Figure C.6), all of which may vary in size and shape and be used in various combinations within a domain. It should be noted that the FVM is not restricted to these four cell types and any *convex polyhedra* can generally be used. Using what is known as a *co-located* scheme, the continuous flow variables $\mathbf{u}$, p, and other scalars are stored at the centroid of each cell (Figure C.7). As was illustrated in the discretization procedure however, the integration over the surface of a control volume is replaced by the summation over the faces of the cell, and in this case the values at the face centroid are required. The face centroid values are interpolated based on the values stored at the centroids of the two neighbouring cells sharing that face, but the important aspect is that because the faces are shared between cells, they are

better thought of entities in their own right, rather than belonging to a cell. In using
what is known as a *face based* data structure, the face, cell, and vertex connectivity
is defined by a *Face* array F and a *Vertex* array V. For the faces f_1 to f_4 in the
portion of a computational grid illustrated in Figure C.7 this would take the form:

$$
F = \begin{vmatrix}
V_1 & V_2 & V_3 & V_4 & , & C_1 & C_2 \\
V_1 & V_5 & V_6 & V_2 & , & C_3 & C_4 \\
V_1 & V_2 & V_7 & V_8 & , & C_1 & C_3 \\
V_1 & V_9 & V_{10} & V_2 & , & C_4 & C_2 \\
& & & . & & & \\
& & & . & & & \\
& & & . & & &
\end{vmatrix}
\qquad
V = \begin{vmatrix}
x_1 & y_1 & z_1 \\
x_2 & y_2 & z_2 \\
x_3 & y_3 & z_3 \\
x_4 & y_4 & z_4 \\
x_5 & y_5 & z_5 \\
& . & \\
& . & \\
& . &
\end{vmatrix}
\qquad (C.45)
$$

where the length of F is the number of faces in the computational grid N_{faces}.
V stores a list of the spatial coordinates of all of the vertices comprised in the
computational grid and the first four entries in a particular row in F each provide a
pointer to a particular row in V, defining the vertices that the given face is comprised
of. Note that the four vertex pointers illustrated in F imply quadrilateral faces,
whereas triangular faces would only require three vertex pointers to define a face.
The remaining two entries in a given row of F are pointers to a cell array C which
identify the two cells sharing the given face. Entries in C store the flow field data
at the cell centroid, and other variables and coefficients associated with the solution
process. In order to solve the linearized discrete transport equation in (C.42), the
connectivity information between neighbouring cells must also be stored explicitly,
and assuming that a given cell c has n neighbours, then the total amount of storage
St required would be:

$$
St = \sum_{c=1}^{N_{cells}} n_c \qquad (C.46)
$$

One convenient method for storage of the cell connectivity information is to let
the cell array C be of length $N_{cells}+1$, and let $C(1) = 1$ and $C(c) = C(c-1)+n_{c-1}$.
Furthermore a neighbour array Nb of length St can be defined, which stores the

pointers back into the cell array C. Therefore, for a given cell c, indices in the cell array C of its neighbouring cells can be found in the Nb array at the locations $Nb(C(c))$ to $Nb(C(c+1))$ [111]. To illustrate via example, for the cells c_1, c_2, c_3 and c_4 in Figure C.7, the portions of the cell array and neighbour array would take the form:

$$C = |1 \quad 5 \quad 9 \quad 13... \quad | \tag{C.47}$$

$$Nb = |3 \quad 2 \quad 5 \quad 12 \quad 4 \quad 7 \quad 6 \quad 1 \quad 10 \quad 4 \quad 1 \quad 11 \quad 9 \quad 8 \quad 2 \quad 3... \quad | \tag{C.48}$$

where it can be seen that the regular spacing between entries in C occurs because each cell has four neighbours. This storage strategy therefore implies that for cell c_1, the indices of its neighbouring cells are located in $Nb(1-4)$ and for cell c_2 in $Nb(5-8)$ and so on.

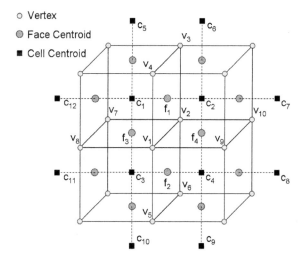

Figure C.7: An example portion of a hexahedral grid, illustrating the relationship between cell centroids, face centroids, and vertices.

Although the end results of the simulation are the pressure and velocity values at the cell centroids of all of the cells comprising the fluid domain, face centroid values are also stored and part of the solution process involves looping over all of the faces in the domain, before performing the loop over all of the cells in a Gauss-Seidel iteration. As can be observed in (C.41) the discretized convective and diffusive terms incorporate the face centroid values of ϕ and $\nabla\phi$ whereas the source term and (as will be explained later) the unsteady term incorporate the cell centroid values of ϕ. All of these terms form parts of the a and b terms for a given cell, but the most logical way to understand their incorporation into the linearized form of (C.42) is to first consider their individual contributions separately and then assemble the overall result.

C.3.1   Evaluation of the Convective Term

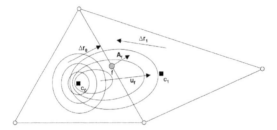

Figure C.8: The effect of convective on the propagation of ϕ through the domain.

Considering first the convective term, the presence of the face value of ϕ necessitates the use of an interpolation scheme to relate the face value to the cell centroid values of neighbouring cells. One method available is to use a *central differencing* scheme, where ϕ_f is taken as the average of the cell centroid values of the two neighbouring cells, another is to use an *upwinding* scheme. To understand the idea behind upwinding, consider the face f, shared by the two cells c_0 and c_1 illustrated in Figure C.8. In the case of pure diffusion (i.e. where the convective term would be zero), ϕ will spread through the domain equally in all directions (shown by the concentric circles in Figure C.8) and is a function of the diffusion coefficient Γ. Since the convective term includes the velocity of the flow $\mathbf{u}$, it effects the transport equations such that the flow variables will spread throughout the domain more strongly in the downstream direction (shown by the ellipses in Figure C.8). This means that the face value of ϕ will be a stronger function of the value stored at the centroid of the upstream cell and it is the upwinding scheme which specifies this function.

While there are numerous upwinding schemes available for evaluating the face value of ϕ, for the present study a *second order* upwind scheme is implemented [3, 111], where the face value of ϕ is interpolated linearly from the upstream cell as:

$$\phi_f = \phi_{c_{up}} + \nabla\phi_{c_{up}} \cdot \Delta\mathbf{r}_{up} \tag{C.49}$$

where $\nabla\phi_{c_{up}}$ is the cell gradient of the upstream cell and Δr_{up} is the vector from the cell centroid to the face centroid. As an aside, the *first order* upwind scheme would set the face value of ϕ to be equal to the cell centroid value of the the upstream cell (i.e. neglecting the second term on the right hand side of (C.49)), but the 'numerical diffusion' introduced by this method tends to 'smear' velocity gradients and may cause a misrepresentation of the flow field in the final solution. By defining the face mass flowrate M_f as $\rho\,\mathbf{u}_f \cdot \mathbf{A}_f$, and using a central difference method to interpolate the face centroid velocity (i.e. approximating the face centroid velocity as the average of the cell centroid velocities of the two neighbouring cells), the convective flux across the face f in Figure C.8) is given as:

$$\rho\,\phi_f\,\mathbf{u}_f \cdot \mathbf{A}_f = max(M_f, 0)\,(\phi_{c_0} + \nabla\phi_{c_0} \cdot \Delta\mathbf{r}_0) + max(-M_f, 0)\,(\phi_{c_1} + \nabla\phi_{c_1} \cdot \Delta\mathbf{r}_1) \tag{C.50}$$

It is important to note that the face area vector $\mathbf{A}_f$ used in the calculation of M_f is directed from c_0 towards c_1, such that the upwind direction can be identified via (C.50). The cell gradients $\nabla\phi_c$ are calculated by making use of the Divergence Theorem:

$$\nabla\phi_c = \frac{1}{V_c} \sum_f^{N_{facenb}} \phi_f \mathbf{A}_f \tag{C.51}$$

where ϕ_f is calculated using a central difference method. If the momentum equation is under consideration, then ϕ_f *is* $\mathbf{u}_f$. This means that computing the face values of ϕ_f and the cell gradients, requires a prior knowledge of $\mathbf{u}_f$, but this is what is being solved for. Furthermore, it is not possible to simply set $\mathbf{u}_f$ equal to ϕ_f, or to directly substitute (C.51) into (C.50) as this would make the resulting system of equations non-linear and therefore unsolvable. The solution to this is to treat $\mathbf{u}_f$ and hence $\nabla\phi_c$ *explicitly*, meaning that the computed values from the previous iteration m are used to compute the face mass flowrates and hence ϕ_f.

When including the contribution of the convective term to the a coefficients for each cell, an initial face loop is performed before a Gauss-Seidel iteration, looping over all of the $Nfaces$ in the domain. During this face loop, the term involving the cell centroid value of ϕ_c is treated implicitly and hence incorporated into the a coefficient, while the implicit treatment of the term involving the cell gradient $\nabla\phi_c$ means that it is incorporated into the b term, such that the overall contributions are:

$$a_{c0} = max\,(M_f, 0) \qquad\qquad a_{c1} = max\,(-M_f, 0) \qquad\qquad \text{(C.52)}$$
$$b_{c0} = max(M_f, 0)\,(\nabla\phi_{c_0}\cdot\Delta\mathbf{r}_0) \qquad b_{c1} = max(-M_f, 0)\,(\nabla\phi_{c_1}\cdot\Delta\mathbf{r}_1) \qquad \text{(C.53)}$$

where the subscripts c_0 and c_1 are the pointers in the Face array F pointing to the two cells in the cell array C that share the face f.

C.3.2   Evaluation of the Diffusive Term

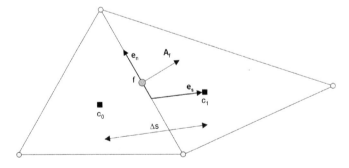

Figure C.9: Two tetrahedral cells (depicted in 2D for simplicity), illustrating the area and unit vectors used in the calculation of the diffusive term.

Evaluation of the diffusive term in the transport equation requires the evaluation of the gradient of ϕ at the face $\nabla\phi_f$, and in order to understand how this is achieved, consider the cell c_0 shown in Figure C.9, illustrated in 2D for simplicity. These cells are considered non-orthogonal since the line joining the cell centroids Δs is not normal to their common face f. This feature complicates the calculation of the diffusive term in that there will be two components to the diffusive flux across a face, one in the direction $\mathbf{e}_s$ between cell centroids, and the other in the direction $\mathbf{e}_n$ in the plane of the face f. To account for both of these terms, and assuming a linear variation in $\nabla\phi$ between cell centroids, the diffusive flux can be defined as:

$$\Gamma_f \nabla\phi_f = D_f \left(\phi_{c_1} - \phi_{c_0}\right) + E_f \tag{C.54}$$

where D_f is known as the *primary diffusive* flux, taking into account the diffusion in the direction of $\mathbf{e}_s$ and defined as:

$$D_f = \frac{\Gamma_f}{\Delta s} \frac{\mathbf{A}_f \cdot \mathbf{A}_f}{\mathbf{A}_f \cdot \mathbf{e}_s} \tag{C.55}$$

and E_f is known as the *secondary diffusive* flux, taking into account the diffusion in the direction of $\mathbf{e}_n$ and defined as:

$$E_f = \Gamma_f \left(\mathbf{A}_f \cdot \nabla\phi_f - \mathbf{e}_s \cdot \nabla\phi_f \frac{\mathbf{A}_f \cdot \mathbf{A}_f}{\mathbf{A}_f \cdot \mathbf{e}_s} \right) \tag{C.56}$$

which is effectively the difference between the total and the primary diffusion. It can be observed in (C.56) that the value of $\nabla\phi_f$ is still required in order to evaluate the secondary diffusion term, but it is calculated in a way that makes the secondary diffusion term explicit in the solver procedure. The gradient of ϕ at the face can be related to the cell centroid gradient values at the two cells sharing the face f as:

$$\nabla\phi_f = \frac{\nabla\phi_{c_0} + \nabla\phi_{c_1}}{2} \tag{C.57}$$

where the cell gradients $\nabla\phi_c$ are calculated by making use of the Divergence Theorem as in (C.51). Using what is known as a *node-based* gradient reconstruction, the face value of ϕ is computed from values at the vertices of the given face [81]. The node based evaluation is known to be more accurate for tetrahedral cells [3] and it is hence this method which is used for the present study. Since ϕ is not stored at the vertices an interpolation is required to determine each vertex value based on the values stored at the centroids of the neighboring cells. For the vertex based evaluation the face value of ϕ is given as:

$$\phi_f = \frac{1}{N_{vertexnb}} \sum_{v}^{N_{vertexnb}} \phi_v \tag{C.58}$$

where ϕ_v is the value of ϕ at a particular vertex v on a face, the total number of which is $N_{vertexnb}$.

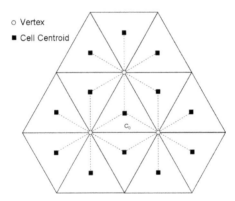

Figure C.10: Tetrahedral cells depicted in 2D for simplicity, illustrating the relation between the vertex values of c_p and the values at the centroids of its neighbors in order to calculate the face value of ϕ.

Consider the cell c_0 in Figure C.10, shown in 2D for simplicity. The nodal values of ϕ are base on a weighted average of the values of ϕ stored at the centers of the neighboring cells which is given as:

$$\phi_v = \frac{\displaystyle\sum_{c}^{N_{cellnb}} w_c \phi_c}{\displaystyle\sum_{c}^{N_{neighbors}} w_c} \tag{C.59}$$

where w_c are the *weighting coefficients*. Determining the weighting coefficients is an optimization problem and for each node it is possible to define the *Laplacian* in the x, y, and z directions [49, 128] as:

$$L\left(x_v\right) = \sum_c^{N_{cellnb}} w_c\left(x_c - x_v\right) \tag{C.60}$$

$$L\left(y_v\right) = \sum_c^{N_{cellnb}} w_c\left(y_c - y_v\right) \tag{C.61}$$

$$L\left(z_v\right) = \sum_c^{N_{cellnb}} w_c\left(z_c - z_v\right) \tag{C.62}$$

where N_{cellnb} is the number of cells neighboring the particular vertex. It is also possible to define a cost function Cs for the weighting functions given as:

$$Cs = \sum_c^{N_{cellnb}} \left(1 - w_c\right)^2 \tag{C.63}$$

because it is desirable to keep the weighting functions as close to 1 as possible. It therefore remains to minimize (C.63) subject to the constraints of (C.60 - C.61) and this can be achieved through the method of *Lagrange multipliers*:

$$Cs = \lambda_x L\left(x_v\right) + \lambda_y L\left(y_v\right) + \lambda_z L\left(z_v\right) \tag{C.64}$$

where λ_x, λ_y and λ_z are the three Lagrange multipliers for three constraints to the optimization problem. Applying the method of Lagrange Multipliers to the gradients for the cost function and the constraints gives:

$$\frac{\partial Cs}{\partial w_{c_i}} = \lambda_x \frac{\partial L\left(x_v\right)}{\partial w_{c_i}} + \lambda_y \frac{\partial L\left(y_v\right)}{\partial w_{c_i}} + \lambda_z \frac{\partial L\left(z_v\right)}{\partial w_{c_i}} \tag{C.65}$$

which provides a system of equations which may be solved for the three Lagrange multipliers to give:

$$\lambda_x = \frac{-I_x\left(I_{yy}I_{zz} - I_{yz}^2\right) + I_x\left(I_{xy}I_{zz} - I_{xz}I_{yz}\right) - I_z\left(I_{xy}I_{yz} - I_{yy}I_{xz}\right)}{I_{xx}\left(I_{yy}I_{zz} - I_{xy}^2\right) - I_{xy}\left(I_{xy}I_{zz} - I_{xz}I_{yz}\right) + I_{xz}\left(I_{xy}I_{yz} - I_{yy}I_{xz}\right)} \tag{C.66}$$

$$\lambda_y = \frac{I_x\left(I_{xy}I_{zz} - I_{xz}I_{yz}\right) - I_y\left(I_{xx}I_{zz} - I_{xz}^2\right) + I_z\left(I_{xx}I_{xy} - I_{xy}I_{xz}\right)}{I_{xx}\left(I_{yy}I_{zz} - I_{xy}^2\right) - I_{xy}\left(I_{xy}I_{zz} - I_{xz}I_{yz}\right) + I_{xz}\left(I_{xy}I_{yz} - I_{yy}I_{xz}\right)} \tag{C.67}$$

$$\lambda_z = \frac{-I_x\left(I_{xy}I_{yz} - I_{yy}I_{xz}\right) + I_y\left(I_{xx}I_{yz} - I_{xy}I_{xz}\right) - I_z\left(I_{xx}I_{yy} - I_{xy}^2\right)}{I_{xx}\left(I_{yy}I_{zz} - I_{xy}^2\right) - I_{xy}\left(I_{xy}I_{zz} - I_{xz}I_{yz}\right) + I_{xz}\left(I_{xy}I_{yz} - I_{yy}I_{xz}\right)} \tag{C.68}$$

where:

$$
\begin{aligned}
I_x &= \sum_c^{N_{neighbors}} (x_c - x_v) & I_{xy} &= \sum_c^{N_{neighbors}} (x_c - x_v)(y_c - y_v) & I_{xx} &= \sum_c^{N_{neighbors}} (x_c - x_v)^2 \\
I_y &= \sum_c^{N_{neighbors}} (y_c - y_v) & I_{yz} &= \sum_c^{N_{neighbors}} (y_c - y_v)(z_c - z_v) & I_{yy} &= \sum_c^{N_{neighbors}} (y_c - y_v)^2 \\
I_z &= \sum_c^{N_{neighbors}} (z_c - z_v) & I_{xz} &= \sum_c^{N_{neighbors}} (x_c - x_v)(z_c - z_v) & I_{zz} &= \sum_c^{N_{neighbors}} (z_c - z_v)^2
\end{aligned}
\tag{C.69}
$$

Finally the weighting coefficients for a given node based on the values at the cell centers of its neighbors are given as:

$$w_c = 1 + \lambda_x\left(x_c - x_v\right) + \lambda_y\left(y_c - y_v\right) + \lambda_z\left(z_c - z_v\right) \tag{C.70}$$

Once the weighting coefficients have been determined the cell gradient $\nabla\phi_c$ may be calculated using (C.51).

The diffusivity Γ_f, is treated explicitly in both the primary and secondary diffusion terms, where the face values are calculated by averaging the cell centroid values between the two cells sharing the face f as:

$$\Gamma_f^{m+1} = \frac{\Gamma_{c_0}^m + \Gamma_{c_1}^m}{2}$$

(C.71)

In terms of the contribution of the diffusive terms to the coefficient a and source term b for a given cell, during the face loop at the beginning of the solution procedure, the primary diffusion term is treated implicitly and hence incorporated into the a coefficient, while the secondary diffusion term is treated explicitly and hence incorporated into the b term as:

$$a_{c_0} = D_f \qquad\qquad a_{c_1} = D_f \qquad (C.72)$$
$$b_{c_0} = -E_f \qquad\qquad b_{c_1} = E_f \qquad (C.73)$$

where again, the subscripts c_0 and c_1 are the pointers in the Face array F, pointing to the two cells in the cell array C, that share the face f. It should be noted that the secondary diffusion term is added to one cell and subtracted from the other, since the diffusive flux leaves one cell and enters the other.

C.3.3 Evaluation of the Source Term

It was shown in Equation C.39 that the source term incorporates both the pressure term and the extra body force term applied in the porous blocks used to represent the cerebrovascular resistance. Since this body force term involves the velocity $\mathbf{u}$, the body force term is treated implicity in the solution procedure. In keeping with the segregated-implicit solution method, the pressure term is treated explicitly. In terms of the contributions to the coefficient a and source term b for a given cell, the two terms are incorporated as:

$$a_c = -\eta \, CVRV_c \tag{C.74}$$

$$b_c = -\sum_f^{N_{facenb}} p_f \mathbf{A}_f \tag{C.75}$$

where it should be noted that in contrast to the convective and diffusive terms, which assigned the contribution to the a and b terms during a loop over all of the faces in the domain, the contribution of the source term is assigned during a *cell* loop which is performed before a Gauss-Seidel iteration.

C.3.4   Evaluation of the Unsteady Term

Up until this point, the explanation of the discretization of the governing equations has considered only the spatial discretization of pressure and velocity. The last remaining term to consider is the unsteady term. Using a *second order backward difference* scheme [3], the unsteady term in (C.39) may be discretized as:

$$\frac{\partial}{\partial t}\int_{CV} \rho \, \phi dV \approx \frac{3\rho\phi_c - 4\rho\phi_c^{t-1} + \rho\phi_c^{t-2}}{2\Delta t} V_c \tag{C.76}$$

where t refers to values of the cell centroid values ϕ_c at previous time steps and Δt is the timestep size. This type of temporal discretization is known as an *implicit time integration* method because it contains the value of ϕ_c at that is being calculated at the current iteration and has the advantage that it is numerically stable for any choice of timestep size. With respect to the contribution of the coefficient a and source term b for a given cell, the value of ϕ at the current timestep is treated implicitly and incorporated into the a coefficient for a given cell, while the terms involving the values of ϕ from previous time steps are already known and hence treated explicitly, and incorporated into the b term such that the overall contribution of the unsteady term is:

$$a_c = \frac{3\rho V_c}{2\Delta t} \tag{C.77}$$

$$b_c = \frac{4\rho V_c \phi^{t-1}}{2\Delta t} - \frac{\rho V_c \phi^{t-2}}{2\Delta t} \tag{C.78}$$

Now that all four terms in the generalized transport equation have been evaluated, the calculation of the a and b terms for a given cell can be delineated as:

$$a_n = \sum_f^{N_{facenb}} max\,(M_f, 0) + D_f \tag{C.79}$$

$$a_c = \sum_n^{N_{cellnb}} a_n - \left(\eta CVR + \frac{3\rho}{2\Delta t}\right) V_c \tag{C.80}$$

$$b_c = \sum_f^{N_{facenb}} max\,(M_f, 0)\,(\nabla\phi_c \cdot \Delta r) + E_f - p_f \mathbf{A}_f + \left(4\phi^{t-1} - \phi^{t-2}\right)\frac{\rho V_c}{2\Delta t} \tag{C.81}$$

where the summation in (C.79) represents the convection and primary diffusion contributions and the summation term in (C.81) represents the secondary diffusion contributions from all of the faces defining a given cell, which are added during the initial face loop. The remaining terms in (C.80) and (C.81) do not involve face values of ϕ and are therefore assigned during a cell loop.

C.3.5 Pressure Velocity Coupling

In solving the discretized momentum equations it is assumed that the pressure values are known. However, the pressure field is not normally known in advance and is obtained as part of the overall solution, so it remains to couple the pressure into the solution of the governing equations. One of the key features of the segregated-implicit solution method is that a single equation is used to solve for a single flow field variable and it has been presented thus far that the three momentum equations are used to solve for the three components of the discrete velocity field. In order to solve for the discrete pressure field the remaining equation, namely the continuity equation, which closes the system of equations, must be transformed into an equation which can be solved for the pressure, known as a *pressure-based* solution method. While there are numerous methods by which this pressure velocity coupling can be performed, for the present study it is achieved using the *Semi Implicit Method for Pressure Linked Equations* or *SIMPLE* pressure velocity coupling method [122]. Before beginning the explanation of the SIMPLE method, some attention must be given to how the discrete cell centroid values of pressure and velocity are related to one another through the momentum equation.

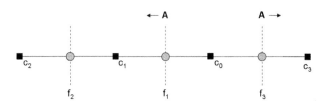

Figure C.11: 1D cells within a fluid domain, illustrating the cell centroids, face centroids and the face Area vectors with respect to the cell c_0.

In order to simplify the discussion, consider the linearized form of the momentum equation applied to a cell c, where the pressure term has been extracted from the source term:

$$a_c \mathbf{u}_c = \sum_{n}^{N_{cellnb}} a_n \mathbf{u}_n + b_c - \sum_{f}^{N_{facenb}} p_f \mathbf{A}_f \tag{C.82}$$

Consider now the computational grid of Figure C.11, where 1D control volumes are presented for the sake of simplicity (although the discussion to follow will apply in higher dimensions), with the cell centroids c_i and cell faces f_i. Assuming that the areas of all faces in the 1D domain are of equal magnitude, and interpolating the face values of pressure as the average of the cell centroid values of the two neighbouring cells, then the pressure term applied to the cell c_0 is evaluated as:

$$\sum_{f}^{N_{facenb}} p_f \mathbf{A}_f = -\mathbf{A}_{f_1} \frac{p_{c_1} + p_{c_0}}{2} + \mathbf{A}_{f_2} \frac{p_{c_0} + p_{c_3}}{2}$$

$$= \frac{\mathbf{A}_{f_1}}{2} \left(p_{c_3} - p_{c_1} \right)$$

(C.83)

It is important to note that the face area vector is always directed away from a control volume or cell, hence the reason for $\mathbf{A}_{f_1}$ being negative. If however, the cell c_1 was under consideration, then $\mathbf{A}_{f_1}$ would be in the opposite direction. Substituting (C.83) into (C.82) and diving by the a_c coefficient yields:

$$\mathbf{u}_{c_0} = \frac{1}{a_{c_0}} \left(\sum_{n}^{N_{cellnb}} a_n \mathbf{u}_n + b_c \right) + \frac{\mathbf{A}_{f_1}}{2 a_{c_0}} \left(p_{c_1} - p_{c_3} \right)$$

$$= \hat{\mathbf{u}}_{c_0} + \frac{\mathbf{A}_{f_1}}{2 a_{c_0}} \left(p_{c_1} - p_{c_3} \right)$$

(C.84)

where the $\hat{\mathbf{u}}_{c_0}$ term includes the effects of neighbouring cells and the source term on the evaluation of the velocity as cell c_0. An important attribute of interpolating the face pressure in this way is that Equation C.84, used to calculate the velocity at a given cell c_0, does not involve the cell centroid pressure p_{c_0}. This can have important consequences as the solution proceeds and gives rise to what is known as a *checkerboard pressure field* (Figure C.12), which is one of the potential problems associated with a co-located scheme. To illustrate via example in the case of cell c_0 in Figure C.11, then the pressure gradient term would be calculated based on the cell centroid pressure of the two neighbouring cells, and the oscillations would be 'seen' as a continuous pressure field. Clearly, this is an un-physical and undesirable result. To circumvent this problem one can either use a *staggered grid* (where pressures are in fact stored at the face centroids, not the cell centroids) or use a different form of interpolation for the face pressure values. Staggered grids are most readily applicable to orthogonal structured grids and so the approach taken here will use the latter option, introducing the interpolation method at a later point in the derivation.

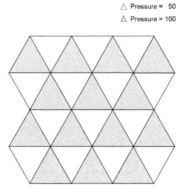

Figure C.12: When using co-located grids, *Checkerboarding* may result, whereby a highly oscillatory and discontinuous pressure field may be obtained. Note: the pressure values shown are arbitrary and only used to illustrate the point

By substitution of the appropriate value of ϕ into (C.41), the discrete form of the continuity equation becomes:

$$\sum_f^{N_{facenb}} \rho\, \mathbf{u}_f \cdot \mathbf{A}_f = \sum_f^{N_{facenb}} M_f = 0 \qquad (C.85)$$

Considering now the face f_1 in Figure C.11, interpolating the face velocity as the average of the cell centroid values of the two neighbouring cells to the face (c_1 and c_0), and using the equation for cell velocity derived in (C.84), the mass flowrate across the face is:

$$
\begin{aligned}
M_f &= \rho \frac{\mathbf{u}_{c_1} + \mathbf{u}_{c_0}}{2} \cdot \mathbf{A}_{f_1} \\
&= \rho \frac{\hat{\mathbf{u}}_{c_1} + \hat{\mathbf{u}}_{c_0}}{2} \cdot \mathbf{A}_{f_1} + \rho \frac{\mathbf{A}_{f_1} \cdot \mathbf{A}_{f_1}}{4 a_{c_1}} \left(p_{c_2} - p_{c_0} \right) + \frac{\mathbf{A}_{f_1} \cdot \mathbf{A}_{f_1}}{4 a_{c_0}} \left(p_{c_1} - p_{c_3} \right)
\end{aligned}
\qquad (C.86)
$$

Returning now to the problem of checkerboarding, the method used to interpolate the pressures is based on the method devised by Rhie and Chow [129] and known as *momentum interpolation* (or *added dissipation*), where an extra pressure term is added into the face mass flowrate calculation of (C.86) such that it becomes:

$$
\begin{aligned}
M_f &= \rho \frac{\hat{\mathbf{u}}_{c_1} + \hat{\mathbf{u}}_{c_0}}{2} \cdot \mathbf{A}_{f_1} + \frac{\mathbf{A}_{f_1} \cdot \mathbf{A}_{f_1}}{4a_{c_1}} (p_{c_2} - p_{c_0}) + \frac{\mathbf{A}_{f_1} \cdot \mathbf{A}_{f_1}}{4a_{c_0}} (p_{c_1} - p_{c_3}) \\
&\quad + \left(\frac{\mathbf{A}_{f_1} \cdot \mathbf{A}_{f_1} (a_{c_0} + a_{c_1})}{2a_{c_0} a_{c_1}} (p_{c_1} - p_{c_0}) - \frac{\mathbf{A}_{f_1} \cdot \mathbf{A}_{f_1}}{4a_{c_1}} (p_{c_2} - p_{c_0}) - \frac{\mathbf{A}_{f_1} \cdot \mathbf{A}_{f_1}}{4a_{c_0}} (p_{c_1} - p_{c_3}) \right) \\
&= \frac{\hat{\mathbf{u}}_{c_1} + \hat{\mathbf{u}}_{c_0}}{2} + \frac{\mathbf{A}_{f_1} \cdot \mathbf{A}_{f_1} (a_{c_0} + a_{c_1})}{2a_{c_0} a_{c_1}} (p_{c_1} - p_{c_0})
\end{aligned}
$$

(C.87)

where the bracketed term corresponds to the added dissipation. Because the linear assumption for the interpolation of face pressures used previously is second order accurate, and furthermore, since it can be shown that the added dissipation term is third order accurate, then its addition into (C.87) does not change the formal second order accuracy of the equation, but does however mean that pressures adjacent to the face under consideration are used, and therefore a checkerboard pressure field could not persist in the final converged solution. When the previous discussion is extended to 3D unstructured grids, the calculation for face mass flowrate becomes:

$$
M_f = \rho \frac{\hat{\mathbf{u}}_{c_1} + \hat{\mathbf{u}}_{c_0}}{2} \cdot \mathbf{A}_{f_1} + \frac{V_{c_0} + V_{c_1}}{a_{c_0} + a_{c_1}} \frac{\mathbf{A}_{f_1} \cdot \mathbf{A}_{f_1}}{\mathbf{A}_{f_1} \cdot \mathbf{e}_s} \left(\frac{\nabla p_{c_0} + \nabla p_{c_1}}{2} + \frac{p_{c_0} - p_{c_1}}{\Delta s} \right) \quad \text{(C.88)}
$$

where ∇p represents the cell pressure gradient, which can be evaluated in the same manner as $\nabla \phi$ in the secondary diffusive term. In addition to the momentum interpolation for face mass flowrate, the face pressure used in the evaluation of the pressure term in (C.82) includes momentum interpolation such that the pressure at a given face f is calculated as:

$$
p_f = \frac{a_{c_0} p_{c_0} + a_{c_1} p_{c_1}}{a_{c_0} + a_{c_1}} \quad \text{(C.89)}
$$

The Gauss-Seidel iteration method begins with an initial 'guess' for both $\mathbf{u}$ and p, however it is unlikely that these will be the correct values for the velocity and pressure field. If the initial guess is denoted by a star $*$, then the correct velocity and pressure field may be denoted:

$$p = p^* + p' \tag{C.90}$$

$$\mathbf{u} = \mathbf{u}^* + \mathbf{u}' \tag{C.91}$$

where the dashed terms are a *correction* to be added to the initial guess, required to achieve the correct pressure and velocity fields. Substituting (C.90) and (C.91) into (C.89) yields an equation for the initial face mass flowrates:

$$M_f^* = \rho \frac{\hat{\mathbf{u}}_{c_1}^* + \hat{\mathbf{u}}_{c_0}^*}{2} \cdot \mathbf{A}_{f_1} + \frac{V_{c_0} + V_{c_1}}{a_{c_0} + a_{c_1}} \frac{\mathbf{A}_f \cdot \mathbf{A}_f}{\mathbf{A}_f \cdot \mathbf{e}_s} \left(\frac{\nabla p_{c_0}^* + \nabla p_{c_1}^*}{2} + \frac{p_{c_0}^* - p_{c_1}^*}{\Delta s} \right) \tag{C.92}$$

and subtracting (C.92) from (C.89) yields an equation for face mass flowrate correction of the form:

$$M_f' = \rho \frac{\hat{\mathbf{u}}_{c_1}' + \hat{\mathbf{u}}_{c_0}'}{2} \cdot \mathbf{A}_{f_1} + \frac{V_{c_0} + V_{c_1}}{a_{c_0} + a_{c_1}} \frac{\mathbf{A}_f \cdot \mathbf{A}_f}{\mathbf{A}_f \cdot \mathbf{e}_s} \left(\frac{\nabla p_{c_0}' + \nabla p_{c_1}'}{2} + \frac{p_{c_0}' - p_{c_1}'}{\Delta s} \right) \tag{C.93}$$

The key feature of the SIMPLE method is that at this point the terms on the right hand side of (C.93), involving the cell neighbour information and the cell pressure gradients, are neglected such that the face mass flowrate correction becomes:

$$M_f' = \frac{V_{c_0} + V_{c_1}}{a_{c_0} + a_{c_1}} \frac{\mathbf{A}_f \cdot \mathbf{A}_f}{\mathbf{A}_f \cdot \mathbf{e}_s} \left(\frac{p_{c_0}' - p_{c_1}'}{\Delta s} \right) \tag{C.94}$$

It is important to note that there is now an explicit relation between pressure correction and mass flowrate correction, and the choice of neglecting the other terms only affects the rate of convergence of the solution, not the final solution itself. Substituting the velocity correction relation of (C.91) into (C.85) the resulting continuity equation takes the form:

$$\sum_{f}^{N_{facenb}} M_f^* + M_f' = 0 \tag{C.95}$$

whereby substituting in the equation for face mass flowrate correction from (C.94) results in an equation for cell pressure correction which may be rearranged to take the form:

$$q_c p_c' = \sum_{n}^{N_{cellnb}} q_n p_n' + r_c \tag{C.96}$$

where:

$$q_n = \frac{V_{c_0} + V_{c_1} \, \mathbf{A}_f \cdot \mathbf{A}_f}{a_{c_0} + a_{c_1} \, \mathbf{A}_f \cdot \mathbf{e}_s} \tag{C.97}$$

$$q_c = \sum_{n}^{N_{cellnb}} q_n \tag{C.98}$$

$$r_c = \sum_{f}^{N_{facenb}} M_f^* \tag{C.99}$$

and it should be noted that r_c represents the mass imbalance in the cell c. Equation C.97 is of the same form as the linearized discrete generalized transport equation and can be solved in the same manner using the point-implicit Gauss-Seidel method to yield the cell pressure correction throughout the domain, which furthermore has the effect of removing the mass imbalance such that the continuity equation is satisfied. In a similar manner to the a and b terms in the generalized transport equation, the q and r terms for each cell are defined via their contributions during a face loop from all of the faces encompassed by a given cell. Once the cell pressure corrections are known face mass flowrates can be updated as:

$$M'_f = M^*_f + \frac{V_{c_0} + V_{c_1}}{a_{c_0} + a_{c_1}} \frac{\mathbf{A}_f \cdot \mathbf{A}_f}{\mathbf{A}_f \cdot \mathbf{e}_s} \left(\frac{p'_{c_0} - p'_{c_1}}{\Delta s} \right) \qquad \text{(C.100)}$$

the cell centroid velocity field can be updated as:

$$\mathbf{u}_c = \mathbf{u}^*_c + \frac{V_c}{a_c} - \sum_f^{N_{neighbours}} \frac{p'_{c_0} + p'_{c_1}}{2} \mathbf{A_f} \qquad \text{(C.101)}$$

and the cell centroid pressure field can be updated as:

$$p_p^{m+1} = p_p^m + \alpha p'_p \qquad \text{(C.102)}$$

where α is again an under-relaxation factor which is included to prevent the pressure from altering too dramatically between iterations.

C.3.6   Monitoring Convergence

As mentioned previously, the Gauss-Seidel methods solves the momentum and pressure correction equations iteratively until a converged solution is obtained. The convergence is judged by monitoring the *residuals*, and the iterations will proceed until the residuals reach a user specified value. If the velocity field at the current iteration is correct then (C.42) will be satisfied, but initially this will not be the case and there will be an imbalance of ϕ in each cell in the domain. The residual for the momentum equations is therefore defined as:

$$R_\phi = \frac{\sum_c^{N_{cells}} \left| \sum_n^{N_{cellnb}} a_n \phi_n + b_c - a_c \phi_c \right|}{\sum_c^{N_{cells}} |a_c \phi_c|} \qquad \text{(C.103)}$$

where the numerator is the summation over all cells in the domain of the magnitude of the imbalance of ϕ within each cell [3]. Monitoring the imbalance of ϕ alone is not particularly useful for judging convergence however, and hence the residuals are scaled with the denominator in (C.103) which can be thought of as a total flowrate of ϕ throughout the domain. The residual for the continuity equation is defined as:

$$R_{continuity} = \frac{\sum_{c}^{N_{cells}} \left| \sum_{f}^{N_{facenb}} \mathbf{u}_f \cdot \mathbf{A}_f \right|}{max \left(\sum_{c}^{N_{cells}} \left| \sum_{f}^{N_{facenb}} \mathbf{u}_f \cdot \mathbf{A}_f \right| \right)_{m=1-5}} \quad \text{(C.104)}$$

where the numerator is the summation over all cells in the domain of the net rate of mass creation within each cell. Again, monitoring the net rate of mass creation alone is not particularly useful for judging convergence and hence the residual is scaled with the maximum rate of mass creation obtained in the first five iterations.

C.3.7   Solver Procedure

Having explained the discretization of all of the terms in the transport equation and outlined the evaluation of the a and b terms for every cell as well as the coupling with the pressure correction equation and the evaluation of the q and r coefficients, the overall solver procedure is illustrated in Figure C.13. The solution begins with an initial guess at the velocity and pressure fields $\mathbf{u}^*$ and p^*. With this initialization a face loop is performed to calculate the face velocities as well as the cell gradients evaluated at each face. Following this loop, a second face loop is performed to add the contributions of convection and primary diffusion to the a and b terms for each cell. Subsequently a cell loop is performed in order to complete the calculation of a_c and b_c for each cell. It is important to note in Figure C.13 that the subscripts c_0 and c_1 represent the contributions to the coefficients of the two cells sharing a particular face f during the face loop, and once the face loop is complete the resulting quantities could be thought of as a_n. The a_c term on the other hand represents the coefficient for a particular cell c within the cell loop and will involve a summation of the a_n coefficients of its neighbours, as well as contributions from the unsteady and source terms. Following the cell loop a Gauss-Seidel iteration is performed sequentially for each of the u, v, w velocity components to obtain the updated velocities $\mathbf{u}_c^*$. A face loop is then performed to calculate the updated face mass flowrates M_f^* (using momentum interpolation), then a second face loop is performed to calculate the contributions of all of the faces to the q and r terms in the pressure correction equation. Following this, a cell loop is performed to complete the calculation of q_c and r_c for each cell and a Gauss Seidel iteration is performed in order to obtain the pressure correction p_c'. A face loop is subsequently performed to update the face mass flowrates, followed by a cell loop to update the cell centroid velocity and pressure fields. The solution is examined for convergence via a comparison of the momentum and continuity residuals. If they are not below the convergence criteria, the updated velocity and pressure fields become the $\mathbf{u}_c^*$ and p_c^* values in the next iteration and the solution process repeats. If the residuals are below the convergence criteria, the updated values of $\mathbf{u}_c$ and p_c are assigned as the velocity and pressure fields for the current timestep and a new timestep is begun.

C.3.8   Boundary Conditions

Up until this point the method used to solve the governing equations has been outlined, but no mention of boundary conditions has been given. Essentially the entire process of solving the governing equations can be thought of as propagating the flow conditions specified at the boundary into the fluid domain. The system of equations used for the present study form what is known as a mixed *parabolic - elliptic* system of equations, meaning that to obtain a solution, boundary conditions must be specified over the *entire* boundary of the fluid domain. Two major classes of boundary conditions exist, known as *Dirichlet* and *Neumann* conditions. Using Dirichlet conditions the value of the flow variable (either $\mathbf{u}$ or p) is specified on the boundary, whereas with Neumann conditions the derivatives of the flow variable are specified on the boundary. For the present study it is the former type of boundary condition which has been implemented, where the no slip condition has been imposed at the arterial wall, and pressures are specified at the afferent and efferent model terminations. In terms of the solution of the discretized governing equations, the no slip condition is implemented by requiring that the face centroid value of $\mathbf{u}$ be zero, such that the convective and diffusive flux contributions to a cell on the boundary face are zero, while the pressure boundary conditions are implemented by assigning the face centroid value of p.

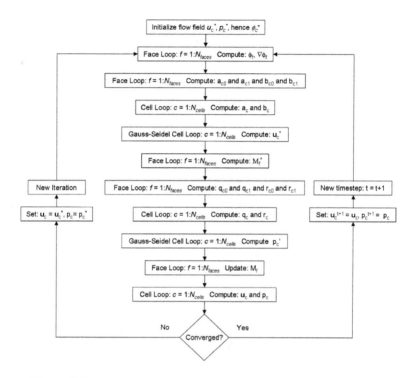

Figure C.13: The procedure used in the solution of the governing equations.

C.3.9   Multigrid Methods

One final issue which needs to be addressed is the use of multigrid methods in the solution of the governing equations. When the Gauss-Seidel iteration method is used to solve the momentum and pressure correction equations, an explicit equation for each cell is obtained relating it to its immediate neighbors. An attractive feature of the Gauss-Seidel method is that updated values of ϕ and p' are spread locally throughout the domain, between a cell and its neighbors relatively quickly, and hence 'high frequency' oscillations in the current solution between neighbouring cells will be rapidly removed. The disadvantage of this method is that the explicit equations imply that a cell is *only* coupled to its immediate neighbours and not to every other cell in the domain, therefore global or 'low frequency' errors spanning over a large number of cells are removed relatively slowly. The implication of this feature of the iterative method is that while the residuals may initially experience a large reduction as the high frequency errors are removed, the solver may then 'stall', where the further reduction associated with the low frequency errors becomes prohibitively slow. Furthermore, as the computational grid is refined to achieve a more physically realistic solution this problem of slow convergence is exacerbated.

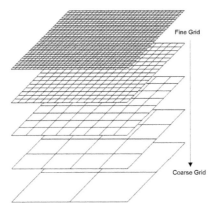

Figure C.14: The multigrid method solves the governing equations on a series of sequentially coarser grids during the iteration process, relating the values computed on the coarse grid, back to the fine grid.

A common solution used to speed up the solver convergence is the use of multigrid methods, where the key concept is that the governing equations are solved on successively coarser grids such that low frequency errors on the original or 'fine' grid appear as high frequency errors on the coarser grids. While the field of multigrid methods is broad, there are two major classifications which can be made: *geometric* and *algebraic* multigrid methods. With the former, groups of cells on the finer grid become a single cell on the coarser grid, and the *a* and *b* terms for each cell on the fine grid are grouped to form a single value for each term on the coarse grid. While the grouping of cells to form a coarser grid can be easily envisaged for structured grids (Figure C.14) the use of geometric multigrid is of less applicability when considering unstructured meshes, as it becomes more difficult to group cells into a single convex polyhedra and the interpolation between grid levels becomes more difficult.

For this reason it is the latter of the two methods incorporated for the present study, namely algebraic multigrid (AMG). The key feature of this method is that the *equations* from a group of cells on the finer level are grouped together to form an a single equation on the coarser level rather than the cells themselves, and this can be achieved with no reference to the underlying geometry, making this method much more applicable to unstructured grids. To understand how the multigrid procedure is achieved, consider the matrix form of the governing equations in (C.43):

$$A\mathbf{x} = \mathbf{b} \tag{C.105}$$

where $\mathbf{x}$ has been used in place of ϕ to illustrate that this method be applied to both the momentum and the pressure correction equations, hence both A and $\mathbf{b}$ can be thought of as representing either equation. It is also important to note that in this case $\mathbf{x}$ represents the solution of the system. However, the solution proceeds with a guessed pressure and velocity field, necessitating the definition of an error term $\mathbf{e} = \mathbf{x} - \mathbf{x}^m$ between the exact solution and the prevailing solution at the iteration m, which can be thought of as the *correction* required to achieve the exact solution. Similarly a *defect* can be defined at the current iteration as $\mathbf{r} = \mathbf{b} - A\mathbf{x}^m$. Substituting these two relations into (C.105) the system becomes:

$$A\mathbf{e} = \mathbf{r} \tag{C.106}$$

The multigrid principal is that the corrections $\mathbf{e}$ at a particular grid level l are estimated from the unknowns $\mathbf{x}$ at a coarser grid level $l + 1$. Solving for corrections on the coarse level requires transferring the defect down from the fine level, (known as *restriction*), computing corrections, and then transferring the corrections back up from the coarse level (known as *prolongation*). To understand how restriction and prolongation are achieved, consider the linearized form of (C.106) applied to a cell c, which may can be written as:

$$a_c e_c = \sum_{n}^{N_{neighbours}} e_n \phi_n + r_c \tag{C.107}$$

Consider now the two cells c_1 and c_2 shown in the computational grid of Figure C.15, which are to be *agglomerated* into a single cell on a coarser grid. Writing out

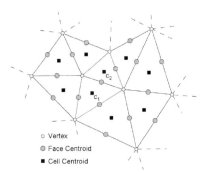

Figure C.15: Two tetrahedral cells (depicted in 2D for simplicity), in the fine level grid, whose equations are to be agglomerated into a single equation on the coarse level grid.

(C.107) for each cell produces the two equations:

$$a^l_{c_1} e^l_{c_1} = a^l_{n_2} e^l_{c_2} + \sum_n^{N_{cellnb}} a^l_n e^l_n + r^l_{c_1} \qquad \text{(C.108)}$$

$$a^l_{c_2} e^l_{c_2} = a^l_{n_1} e^l_{c_1} + \sum_n^{N_{cellnb}} a^l_n e^l_n + r^l_{c_2} \qquad \text{(C.109)}$$

where the relationship between the two cells has been separated from the neighbour summation term. Using a *piecewise constant interpolation* method [70] the correction obtained on the coarse level grid is applied to both cells $e^l_{c_1}$ and $e^l_{c_2}$ on the fine level such that $x^{l+1}_{c_1} = e^l_{c_1} = e^l_{c_2}$. Substituting this relation into (C.108) and (C.109) and rearranging produces the coarse level equation:

$$\sum_{k=1}^{2} \left(a^l_{c_k} + a^l_{n_k} \right) x^{l+1}_{c_1} = \sum_n^{N_{cellnb}} a^l_n x^{l+1}_n + \sum_{k=1}^{2} r^l_{c_k} \qquad \text{(C.110)}$$

where the summation term for $x^{l+1}_{c_1}$ can be thought of as $a^{l+1}_{c_1}$ and the neighbour summation term involves all of the neighbours of cells c_1 and c_2 at the level l. It can also be observed that the defect on the coarse level grid is the summation of the defects on the fine level grid. One method for the *agglomeration* of fine level equations into coarse level equations involves associating a *coarse index* with each fine level cell and initializing it to zero. Furthermore a coarse level cell counter n can be defined and set to 1. By then looping over all of the equations in the current level, if an equation has not been grouped (i.e. its coarse index is zero), group it and its neighbour for which the a_c coefficient is the largest, assign them the coarse level index n, and increment n by 1. In this manner, at the end of the loop, n will be equal to the number of coarse level cells which may subsequently be looped over in a Gauss-Seidel iteration. Agglomerating cells based on the value of a_c helps propagate information into the computational domain more rapidly, providing one of the attractive features of AMG.

Generally the use of AMG in the solution procedure involves a number of multigrid cycles where the computational grid is recursively refined. At each grid level, two Gauss-Seidel *sweeps* are performed, where a sweep is defined as looping through the cells from 1 to N_{cells}, computing the updated values, and then looping in the reverse direction from N_{cells} to 1 computing the updated values so as to avoid bias in the iteration procedure. The three components of a multigrid cycle involve a pre-relaxation sweep where a Gauss-Seidel iteration is performed, transferring the defect down to a coarser grid, transferring the correction back to the finer grid, and performing a post-relaxation sweep, where another Gauss-Seidel iteration is performed. For the pressure correction equation the multigrid cycle performed is known as the *V cycle*, where no pre-relaxation sweeps were performed, one post-relaxation sweep is performed, and the maximum number of coarsening levels was set to 20, coarsening by two cells each time. With these parameters a single Gauss-Seidel iteration (with reference to the solver procedure in Figure C.13) can be thought of as recursively coarsening the grid twenty times, performing a Gauss-Seidel iteration on the coarsest level, transferring the correction back to the next finest level, performing another Gauss-Seidel iteration, with the process repeating until the original finest grid is encountered. For the momentum equations a multigrid cycle known as the *Flexible cycle* was used. In this case, rather than utilizing a fixed recursive procedure, the coarser grid calculations are invoked when the rate of residual reduction on the current grid level is too slow. Furthermore, the transition to the finer grid is made when the iterative solution of the correction on the current coarse grid level is sufficiently converged and should thus be applied to the solution on the next finest level.

References

[1]

[2] emedicine. *http://www.emedicine.com.*

[3] Fluent 6.3 user's guide. *http://www.fluentusers.com.*

[4] Interactive course about mri physics. *http://www.e-mri.org/.*

[5] Internet stroke centre. *http://www.strokecenter.org.*

[6] Revise mri - animated tutorials. *http://www.revisemri.com/tutorials/.*

[7] Aalkjaer, C. and L. Poston (1996). Effects of ph on vascular tension: Which are the important mechanisms? *Journal of Vascular Research 33*(5), 347–359.

[8] Adams, H. P., B. H. Bendixen, L. J. Kappelle, J. Biller, B. B. Love, D. L. Gordon, and E. E. Marsh (1993). Classification of subtype of acute ischemic stroke - definitions for use in a multicenter clinical-trial. *Stroke 24*(1), 35–41.

[9] Alastruey, J., S. Moore, K. Parker, J. Peiro, T. David, and S. Sherwin (2007). Reduced modelling of blood flow in the cerebral circulation: Coupling 1-d, 0-d and cerebral auto-regulation models. *Int. Jour. Num. Methods Fluids 56*, 1061–1067.

[10] Alastruey, J., K. Parker, J. Peiro, S. M. Byrd, and S. Sherwin (2007). Modelling the circle of willis to assess the effects of anatomical variations and occlusions on cerebral flows. *Journal of Biomechanics 40*, 1794–1805.

[11] Alpers, B. J. and R. G. Berry (1963). Circle of willis in cerebral vascular disorders. *Archives of Neurology 8*(1), 398–402.

[12] Alpers, B. J., R. G. Berry, and R. M. Paddison (1959a). Anatomical studies of the circle of willis in normal brain. *Archives of Neurology and Psychiatry 81*(4), 409–418.

[13] Alpers, B. J., R. G. Berry, and R. M. Paddison (1959b). Anatomical studies of the circle of willis in normal brain. *AMA Arch Neurol Psychiatry 81*(4), 409–18.

[14] Archer, S. L., J. M. C. Huang, V. Hampl, D. P. Nelson, P. J. Shultz, and E. K. Weir (1994). Nitric-oxide and cgmp cause vasorelaxation by activation of a charybdotoxin-sensitive k-channel by cgmp-dependent protein-kinase. *Proceedings of the National Academy of Sciences of the United States of America 91*(16), 7583–7587.

[15] Austin, C. and S. Wray (2000). Interactions between ca2+ and h+ and functional consequences in vascular smooth muscle. *Circulation Research 86*(3), 355–363.

[16] Banaji, M., A. Tachtsidis, D. Delpy, and S. Baigent (2005a). A physiological model of cerebral blood flow control. *Mathematical Biosciences 194*(2), 125–173.

[17] Banaji, M., A. Tachtsidis, D. Delpy, and S. Baigent (2005b). A physiological model of cerebral blood flow control. *Mathematical Biosciences 194*(2), 125–173.

[18] Baron, J. C. (1999). Mapping the ischaemic penumbra with pet: Implications for acute stroke treatment. *Cerebrovascular Diseases 9*(4), 193–201.

[19] Baron, J. C., M. G. Bousser, A. Rey, A. Guillard, D. Comar, and P. Castaigne (1981). Reversal of focal misery-perfusion syndrome by extra-intracranial arterial bypass in hemodynamic cerebral-ischemia - a case-study with o-15 positron emission tomography. *Stroke 12*(4), 454–459.

[20] Baukrowitz, T. and B. Fakler (2000). K-atp channels gated by intracellular nucleotides and phospholipids. *European Journal of Biochemistry 267*(19), 5842–5848.

[21] Bogousslavsky, J., G. Vanmelle, and F. Regli (1988). The lausanne stroke registry - analysis of 1,000 consecutive patients with 1st stroke. *Stroke 19*(9), 1083–1092.

[22] Bolz, S. S., L. Vogel, D. Sollinger, R. Derwand, C. de Wit, G. Loirand, and U. Pohl (2003). Nitric oxide-induced decrease in calcium sensitivity of resistance arteries is attributable to activation of the myosin light chain phosphatase and antagonized by the rhoa/rho kinase pathway. *Circulation 107*(24), 3081–3087.

[23] Brian, J. E., D. D. Heistad, and F. M. Faraci (1995). Dilatation of cerebral arterioles in response to lipopolysaccharide in-vivo. *Stroke 26*(2), 277–280.

[24] Buijs, P. C., M. J. Krabbe-Hartkamp, C. J. G. Bakker, E. E. de Lange, L. M. P. Ramos, M. M. B. Breteler, and W. P. T. M. Mali (1998). Effect of age on cerebral blood flow: Measurement with ungated two-dimensional phase-contrast mr angiography in 250 adults. *Radiology 209*(3), 667–674.

[25] Cady, E. B., A. Chu, A. M. D. Costello, D. T. Delpy, R. M. Gardiner, P. L. Hope, and E. O. R. Reynolds (1987). Brain intracellular ph and metabolism during hypercapnia and hypocapnia in the newborn lamb. *Journal of Physiology-London 382*, 1–14.

[26] Caplan, L. R. and M. Hennerici (1998). Impaired clearance of emboli (washout) is an important link between hypoperfusion, embolism, and ischemic stroke. *Archives of Neurology 55*(11), 1475–1482.

[27] Cassot, F., V. Vergeur, P. Bossuet, B. Hillen, M. Zagzoule, and J. P. Marcvergnes (1995). Effects of anterior communicating artery diameter on cerebral hemodynamics in internal carotid-artery disease - a model study. *Circulation 92*(10), 3122–3131.

[28] Cassot, F., M. Zagzoule, and J. P. Marc-Vergnes (2000). Hemodynamic role of the circle of willis in stenoses of internal carotid arteries. an analytical solution of a linear model. *Journal of Biomechanics 33*(4), 395–405.

[29] Cebral, J., Lohner, P. J. Yim, and J. E. Burgess (2001). Blood flow predictions during neuro-surgery and carotid artery stenting. *International Journal of Bioelectromagnetism 3*(2), 1–12.

[30] Cebral, J., R. Lohner, P. L. Choyke, and P. J. Yim (2002). Parallel patient-specific computational haemodynamics. *Applied Parallel Computing 2367*, 18–34.

[31] Cebral, J. R., M. A. Castro, O. Soto, R. Lohner, and N. Alperin (2003). Blood-flow models of the circle of willis from magnetic resonance data. *Journal of Engineering Mathematics 47*, 369–386.

[32] Charbel, F. T., M. D. Zhao, S. Amin-Hanjani, W. Hoffman, X. J. Du, and M. E. Clark (2004). A patient-specific computer model to predict outcomes of the balloon occlusion test. *Journal of Neurosurgery 101*(6), 977–988.

[33] Collins, R. E. (1961). *Flow of Fluids through Porous Materials*. New York: Reinhold Publishing Corp.

[34] Costa, F. and I. Biaggioni (1998). Role of nitric oxide in adenosine-induced vasodilation in humans. *Hypertension 31*(5), 1061–1064.

[35] Dash, R. K. and J. B. Bassingthwaighte (2004). Blood hbo(2) and hbco(2) dissociation curves at varied o-2, co2, ph, 2,3-dpg and temperature levels. *Annals of Biomedical Engineering 32*(12), 1676–1693.

[36] David, T., M. Brown, and A. Ferrandez (2003a). Auto-regulation and blood flow in the cerebral circulation. *International Journal for Numerical Methods in Fluids 43*, 701–713.

[37] David, T., M. Brown, and A. Ferrandez (2003b). Auto-regulation and blood flow in the cerebral circulation. *International Journal for Numerical Methods in Fluids 43*(6-7), 701–713.

[38] Davignon, J. and P. Ganz (2004). Role of endothelial dysfunction in atherosclerosis. *Circulation 109*(23), 27–32.

[39] Davis, M. J. and M. A. Hill (1999). Signaling mechanisms underlying the vascular myogenic response. *Physiological Reviews 79*(2), 387–423.

[40] Derdeyn, C. P., R. L. Grubb, and W. J. Powers (1999). Cerebral hemodynamic impairment - methods of measurement and association with stroke risk. *Neurology 53*(2), 251–259.

[41] Derdeyn, C. P., T. O. Videen, K. D. Yundt, S. M. Fritsch, D. A. Carpenter, R. L. Grubb, and W. J. Powers (2002). Variability of cerebral blood volume and oxygen extraction: stages of cerebral haemodynamic impairment revisited. *Brain 125*, 595–607.

[42] Derdeyn, C. P., K. D. Yundt, T. O. Videen, D. A. Carpenter, R. L. Grubb, and W. J. Powers (1998). Increased oxygen extraction fraction is associated with prior ischemic events in patients with carotid occlusion. *Stroke 29*(4), 754–758.

[43] Enzmann, D. R., M. R. Ross, M. P. Marks, and N. J. Pelc (1994). Blood-flow in major cerebral-arteries measured by phase-contrast cine mr. *American Journal of Neuroradiology 15*(1), 123–129.

[44] Faraci, F. M. and D. D. Heistad (1998). Regulation of the cerebral circulation: Role of endothelium and potassium channels. *Physiological Reviews 78*(1), 53–97.

[45] Ferrandez, A., T. David, and M. D. Brown (2002a). Numerical models of auto-regulation and blood flow in the cerebral circulation. *Comput Methods Biomech Biomed Engin 5*(1), 7–19.

[46] Ferrandez, A., T. David, and M. D. Brown (2002b). Numerical models of auto-regulation and blood flow in the cerebral circulation. *Comput Methods Biomech Biomed Engin 5*(1), 7–19.

[47] Fleming, I. and R. Busse (1999). No - the primary edrf. *Journal of Molecular and Cellular Cardiology 31*(1), 5–14.

[48] Frangi, A. F., W. J. Niessen, K. L. Vincken, and M. A. Viergever (1998). Multiscale vessel enhancement filtering. *Medical Image Computing and Computer-Assisted Intervention - Miccai'98 1496*, 130–137.

[49] Frink, N. T. (1994). Recent progress toward a three-dimensional unstructured navier-stokes flow solver. *AIAA-94-0061*.

[50] Gibbs, J. M., K. L. Leenders, R. J. S. Wise, and T. Jones (1984). Evaluation of cerebral perfusion reserve in patients with carotid-artery occlusion. *Lancet 1*(8370), 182–186.

[51] Gibo, H., C. C. Carver, A. L. Rhoton, C. Lenkey, and R. J. Mitchell (1981). Microsurgical anatomy of the middle cerebral-artery. *Journal of Neurosurgery 54*(2), 151–169.

[52] Gijsen, F. J., E. Allanic, F. N. van de Vosse, and J. D. Janssen. The influence of the non-newtonian properties of blood on the flow in large arteries: unsteady flow in a 90 degrees curved tube. *J Biomech 32*(7), 705–13. 0021-9290 Year = 1999.

[53] Gijsen, F. J., F. N. van de Vosse, and J. D. Janssen (1999). The influence of the non-newtonian properties of blood on the flow in large arteries: steady flow in a carotid bifurcation model. *J Biomech 32*(6), 601–8.

[54] Gutierrez, G. (2004). A mathematical model of tissue-blood carbon dioxide exchange during hypoxia. *American Journal of Respiratory and Critical Care Medicine 169*(4), 525–533.

[55] Guyton, A. C. and J. E. Hall (2000). *Textbook of medical physiology* (10th ed.). Philadelphia: Saunders.

[56] Haacke, E. M. (1999). *Magnetic resonance imaging : physical principles and sequence design*. New York: J. Wiley and Sons.

[57] Hairer, E., C. Lubich, and M. Roche (1989). The numerical-solution of differential-algebraic systems by runge-kutta methods. *Lecture Notes in Mathematics 1409*, 1–137.

[58] Hartkamp, M. J., J. van der Grond, F. E. de Leeuw, J. C. de Groot, A. Algra, B. Hillen, M. M. B. Breteler, and W. P. T. M. Mali (1998). Circle of willis: Morphologic variation on three-dimensional time-of-flight mr angiograms. *Radiology 207*(1), 103–111.

[59] Henrion, D. (2005). Pressure and flow-dependent tone in resistance arteries - role of myogenic tone. *Archives Des Maladies Du Coeur Et Des Vaisseaux 98*(9), 913–921.

[60] Hill, M. A., M. J. Davis, G. A. Meininger, S. J. Potocnik, and T. V. Murphy (2006). Arteriolar myogenic signalling mechanisms: Implications for local vascular function. *Clinical Hemorheology and Microcirculation 34*(1-2), 67–79.

[61] Hillen, B., T. Gaasbeek, and H. W. Hoogstraten (1982). A mathematical-model of the flow in the posterior communicating arteries. *Journal of Biomechanics 15*(6), 441–449.

[62] Hillen, B., H. W. Hoogstraten, and L. Post (1986a). A mathematical-model of the flow in the circle of willis. *Journal of Biomechanics 19*(3), 187–194.

[63] Hillen, B., H. W. Hoogstraten, and L. Post (1986b). A mathematical model of the flow in the circle of willis. *J Biomech 19*(3), 187–94.

[64] Horiuchi, T., H. H. Dietrich, K. Hongo, T. Goto, and R. G. Dacey (2002). Role of endothelial nitric oxide and smooth muscle potassium channels in cerebral arteriolar dilation in response to acidosis. *Stroke 33*(3), 844–849.

[65] Horiuchi, T., H. H. Dietrich, S. Tsugane, and R. G. Dacey (2001). Role of potassium channels in regulation of brain arteriolar tone - comparison of cerebrum versus brain stem. *Stroke 32*(1), 218–224.

[66] Hornak, J. P. (1999). Teaching nmr using online textbooks. *Molecules 4*(12), 353–365.

[67] Hudetz, A. G., J. H. Halsey, C. R. Horton, K. A. Conger, and D. D. Reneau (1982). Mathematical simulation of cerebral blood-flow in focal ischemia. *Stroke 13*(5), 693–700.

[68] Hughes, W. F. and J. A. Brighton (1967). *Shaum's Outline of Theory and Problems of Fluid Dynamics.* New York: McGraw-Hill.

[69] Hundley, W. G., D. W. Kitzman, T. M. Morgan, C. A. Hamilton, S. N. Darty, K. P. Stewart, D. M. Herrington, K. M. Link, and W. C. Little (2001). Cardiac cycle-dependent changes in aortic area and distensibility are reduced in older patients with isolated diastolic heart failure and correlate with exercise intolerance. *Journal of the American College of Cardiology 38*(3), 796–802.

[70] Hutchinson, B. R. and G. D. Raithby (1986). A multigrid method based on the additive correction strategy. *Numerical Heat Transfer 9*, 511–537.

[71] Iadecola, C., F. Y. Zhang, and X. H. Xu (1994). Sin-1 reverses attenuation of hypercapnic cerebrovasodilation by nitric-oxide synthase inhibitors. *American Journal of Physiology 267*(1), R228–R235.

[72] Ishizaka, H. and L. Kuo (1996). Acidosis-induced coronary arteriolar dilation is mediated by atp-sensitive potassium channels in vascular smooth muscle. *Circulation Research 78*(1), 50–57.

[73] Jung, A., R. Faltermeier, R. Rothoerl, and A. Brawanski (2005a). A mathematical model of cerebral circulation and oxygen supply. *Journal of Mathematical Biology 51*(5), 491–507.

[74] Jung, A., R. Faltermeier, R. Rothoerl, and A. Brawanski (2005b). A mathematical model of cerebral circulation and oxygen supply. *Journal of Mathematical Biology 51*(5), 491–507.

[75] Kanno, I., K. Uemura, S. Higano, M. Murakami, H. Iida, S. Miura, F. Shishido, A. Inugami, and I. Sayama (1988). Oxygen extraction fraction at maximally vasodilated tissue in the ischemic brain estimated from the regional carbon dioxide responsiveness measured by positron emission tomography. *Journal of Cerebral Blood Flow and Metabolism 8*(2), 227–235.

[76] Karch, R., F. Neumann, M. Neumann, and W. Schreiner (2000). Staged growth of optimized arterial model trees. *Annals of Biomedical Engineering 28*(5), 495–511.

[77] Keener, J. P. and J. Sneyd (1998). *Mathematical physiology*. Interdisciplinary applied mathematics ; v. 8. New York: Springer.

[78] Kellyhayes, M., P. A. Wolf, G. E. Gresham, and R. B. Dagnostino (1988). Course of recovery following stroke - the framingham-study. *Archives of Physical Medicine and Rehabilitation 69*(9), 736–736.

[79] Kim, C. S., C. Kiris, and D. Kwak (2004). Numerical models of human circulatory system under altered gravity: Brain circulation. *AIAA 2004-1092*, 1–12.

[80] Kim, C. S. S., C. Kiris, D. Kwak, and T. David (2006). Numerical simulation of local blood flow in the carotid and cerebral arteries under altered gravity. *Journal of Biomechanical Engineering-Transactions of the Asme 128*(2), 194–202.

[81] Kim, S. E., B. Makarov, and D. Caraeni (2003). A multi-dimensional linear reconstruction scheme for arbitrary unstructured grids. *Fluent Technical Notes* (TN210), 1–23.

[82] Kim, S. E., M. S. R., J. Y. Murthy, and C. D. (1997). A reynolds-averaged navier-stokes solver using an unstructured mesh based finit-volume scheme. *Fluent Technical Notes* (TN117), 1–22.

[83] Kleinstreuer, C. (1997). *Engineering fluid dynamics : an interdisciplinary systems approach*. Cambridge: Cambridge University Press.

[84] Koller, R. L. (1982). Recurrent embolic cerebral infarction and anticoagulation. *Neurology 32*(3), 283–285.

[85] Komiyama, M., H. Nakajima, M. Nishikawa, and T. Yasui (1998). Middle cerebral artery variations: Duplicated and accessory arteries. *American Journal of Neuroradiology 19*(1), 45–49.

[86] Kontos, H. A., A. J. Raper, and J. L. Patterson (1977). Analysis of vasoactivity of local ph, pco2 and bicarbonate on pial vessels. *Stroke 8*(3), 358–360.

[87] Lagaud, G., V. Karicheti, H. J. Knot, G. J. Christ, and I. Laher (2002). Inhibitors of gap junctions attenuate myogenic tone in cerebral arteries. *American Journal of Physiology-Heart and Circulatory Physiology 283*(6), H2177–H2186.

[88] Leenders, K. L., D. Perani, A. A. Lammertsma, J. D. Heather, P. Buckingham, M. J. R. Healy, J. M. Gibbs, R. J. S. Wise, J. Hatazawa, S. Herold, R. P. Beaney, D. J. Brooks, T. Spinks, C. Rhodes, R. S. J. Frackowiak, and T. Jones (1990).

Cerebral blood-flow, blood-volume and oxygen utilization - normal values and effect of age. *Brain 113*, 27–47.

[89] Lindauer, U., J. Vogt, S. Schuh-Hofer, J. P. Dreier, and U. Dirnagl (2003). Cerebrovascular vasodilation to extraluminal acidosis occurs via combined activation of atp-sensitive and ca2+-activated potassium channels. *Journal of Cerebral Blood Flow and Metabolism 23*(10), 1227–1238.

[90] Lodi, C. A., A. Ter Minassian, L. Beydon, and M. Ursino (1998). Modeling cerebral autoregulation and co2 reactivity in patients with severe head injury. *American Journal of Physiology-Heart and Circulatory Physiology 43*(5), H1729–H1741.

[91] Lodi, C. A. and M. Ursino (1999). Hemodynamic effect of cerebral vasospasm in humans - a modeling study. *Annals of Biomedical Engineering 27*(2), 257–273.

[92] Lorensen, W. E. and C. H. E. (1987). Marching cubes: A high resolution 3d surface contruction algorithm. *Computer Graphics 21*(4), 163–169.

[93] Lotz, J., C. Meier, A. Leppert, and M. Galanski (2002). Cardiovascular flow measurement with phase-contrast mr imaging: Basic facts and implementation. *Radiographics 22*, 651–671.

[94] Lu, K., J. W. Clark, F. H. Ghorbel, C. S. Robertson, D. L. Ware, J. B. Zwischenberger, and A. Bidani (2004). Cerebral autoregulation and gas exchange studied using a human cardiopulmonary model. *American Journal of Physiology-Heart and Circulatory Physiology 286*(2), H584–H601.

[95] Lumb, A. B. and J. F. Nunn (2005). *Nunn's applied respiratory physiology* (6th ed.). Oxford ; Philadelphia: Elsevier Butterworth Heinemann.

[96] Magosso, E. and M. Ursino (2004). Modelling study of the acute cardiovascular response to hypocapnic hypoxia in healthy and anaemic subjects. *Medical and Biological Engineering and Computing 42*(2), 158–166.

[97] Marks, M. P., N. J. Pelc, M. R. Ross, and D. R. Enzmann (1992). Determination of cerebral blood-flow with a phase-contrast cine mr imaging technique - evaluation of normal subjects and patients with arteriovenous-malformations. *Radiology 182*(2), 467–476.

[98] Matsumoto, N., J. P. Whisnant, L. T. Kurland, and H. Okazaki (1973). Natural history of stroke in rochester, minnesota, 1955 through 1969: An extension of a previous study, 1945 through 1954. *Stroke 4*(1), 20–29.

[99] Mchenry, L. C., J. F. Fazekas, and J. F. Sullivan (1961). Cerebral hemodynamics of syncope. *American Journal of the Medical Sciences 241*(2), 173–178.

[100] McRobbie, D. W. (2003). *MRI From Picture to Proton.* New York: Cambridge.

[101] Middleman, S. (1972). *Transport phenomena in the cardiovascular system.* New York,: Wiley-Interscience.

[102] Mitsis, G. D., P. N. Ainslie, M. J. Poulin, P. A. Robbins, and V. Z. Marmarelis (2004). Nonlinear modeling of the dynamic effects of arterial pressure and blood gas variations on cerebral blood flow in healthy humans. *Post-Genomic Perspectives in Modeling and Control of Breathing 551*, 259–265.

[103] Mitsis, G. D. and V. Z. Marmarelis (2002). Modeling of nonlinear physiological systems with fast and slow dynamics - 1 methodology. *Annals of Biomedical Engineering 30*(2), 272–281.

[104] Mitsis, G. D., M. J. Poulin, P. A. Robbins, and V. Z. Marmarelis (2004). Nonlinear modeling of the dynamic effects of arterial pressure and co2 variations on cerebral blood flow in healthy humans. *Ieee Transactions on Biomedical Engineering 51*(11), 1932–1943.

[105] Mitsis, G. D., R. Zhang, B. D. Levine, and V. Z. Marmarelis (2002). Modeling of nonlinear physiological systems with fast and slow dynamics - ii application to cerebral autoregulation. *Annals of Biomedical Engineering 30*(4), 555–565.

[106] Mitsis, G. D., R. Zhang, B. D. Levine, and V. Z. Marmarelis (2006). Cerebral hemodynamics during orthostatic stress assessed by nonlinear modeling. *Journal of Applied Physiology 101*(1), 354–366.

[107] Momjian-Mayor, I. and J. C. Baron (2005). The pathophysiology of watershed infarction in internal carotid artery disease - review of cerebral perfusion studies. *Stroke 36*(3), 567–577.

[108] Moncada, S. and E. A. Higgs (2006). The discovery of nitric oxide and its role in vascular biology. *British Journal of Pharmacology 147*, S193–S201.

[109] Moore, S. and T. David (2006). Auto-regulated blood flow in the cerebral-vasculature. *Journal of Biomechanical Science and Engineering 1*(1), 1–14.

[110] Moore, S., T. David, J. G. Chase, J. Arnold, and J. Fink (2006). 3d models of blood flow in the cerebral vasculature. *Journal of Biomechanics 39*(8), 1454–1463.

[111] Murthy, J. Y. (2002). Numerical mehods in heat, mass, and momentum transfer.

[112] Newell, D. W., R. Aaslid, A. Lam, T. S. Mayberg, and H. R. Winn (1994). Comparison of flow and velocity during dynamic autoregulation testing in humans. *Stroke 25*(4), 793–797.

[113] Nielson, G. M. and B. Hamann (1991). The asymptotic decider: Resolving the ambiguity in marching cubes. *Proc Vis 91*, 83.

[114] Olufsen, M. S., A. Nadim, and L. A. Lipsitz (2002a). Dynamics of cerebral blood flow regulation explained using a lumped parameter model. *American Journal of Physiology-Regulatory Integrative and Comparative Physiology 282*(2), R611–R622.

[115] Olufsen, M. S., A. Nadim, and L. A. Lipsitz (2002b). Dynamics of cerebral blood flow regulation explained using a lumped parameter model. *Am J Physiol Regul Integr Comp Physiol 282*(2), R611–22.

[116] Olufsen, M. S., J. T. Ottesen, H. T. Tran, L. M. Ellwein, L. A. Lipsitz, and V. Novak (2005). Blood pressure and blood flow variation during postural change from sitting to standing - model development and validation. *Journal of Applied Physiology 99*(4), 1523–1537.

[117] Panerai, R. B., B. J. Carey, and J. F. Potter (2003). Short-term variability of cerebral blood flow velocity responses to arterial blood pressure transients. *Ultrasound in Medicine and Biology 29*(1), 31–38.

[118] Panerai, R. B., M. Chacon, R. Pereira, and D. H. Evans (2004). Neural network modelling of dynamic cerebral autoregulation - assessment and comparison with established methods. *Medical Engineering and Physics 26*(1), 43–52.

[119] Panerai, R. B., S. L. Dawson, and J. F. Potter (1999). Linear and nonlinear analysis of human dynamic cerebral autoregulation. *American Journal of Physiology-Heart and Circulatory Physiology 277*(3), H1089–H1099.

[120] Panerai, R. B., A. W. R. Kelsall, J. M. Rennie, and D. H. Evans (1996). Analysis of cerebral blood flow autoregulation in neonates. *Ieee Transactions on Biomedical Engineering 43*(8), 779–788.

[121] Panerai, R. B., J. M. Rennie, A. W. R. Kelsall, and D. H. Evans (1998). Frequency-domain analysis of cerebral autoregulation from spontaneous fluctuations in arterial blood pressure. *Medical and Biological Engineering and Computing 36*(3), 315–322.

[122] Patankar, S. V. (1980). *Numerical heat transfer and fluid flow*. Series in computational methods in mechanics and thermal sciences. s.l.: Taylor and Francis.

[123] Peng, H. L., P. E. Jensen, H. Nilsson, and C. Aalkjaer (1998). Effect of acidosis on tension and ca2+i in rat cerebral arteries - is there a role for membrane potential. *American Journal of Physiology-Heart and Circulatory Physiology 43*(2), H655–H662.

[124] Perlmutter, D. and A. L. Rhoton (1978). Microsurgical anatomy of distal anterior cerebral-artery. *Journal of Neurosurgery 49*(2), 204–228.

[125] Pollanen, M. S. and J. H. N. Deck (1990). The mechanism of embolic watershed infarction - experimental studies. *Canadian Journal of Neurological Sciences 17*(4), 395–398.

[126] Powers, W. J., M. E. Raichle, and R. L. Grubb (1985). Positron emission tomography to assess cerebral perfusion. *Lancet 1*(8420), 102–103.

[127] Ramsay, S. C., K. Murphy, S. A. Shea, K. J. Friston, A. A. Lammertsma, J. C. Clark, L. Adams, A. Guz, and R. S. J. Frackowiak (1993). Changes in global cerebral blood-flow in humans - effect on regional cerebral blood-flow during a neural activation task. *Journal of Physiology-London 471*, 521–534.

[128] Rausch, R. D., J. T. Batina, and H. T. Y. Yang (1991). Spatial adaption procedure on unstructured meshes for accurate unsteady aerodynamic flow computation. *AIAA-91-1106-CP*, 1904–18.

[129] Rhie, C. M. and W. L. Chow (1983). Numerical study of the turbulent-flow past an airfoil with trailing edge separation. *Aiaa Journal 21*(11), 1525–1532.

[130] Sacco, S. E., J. P. Whisnant, J. P. Broderick, S. J. Phillips, and W. M. Ofallon (1991). Epidemiologic characteristics of lacunar infarcts in a population. *Stroke 22*(10), 1236–1241.

[131] Sato, Y., S. Nakajima, H. Atsumi, T. Koller, G. Gerig, S. Yoshida, and R. Kiki-
nis (1997). 3d multi-scale line filter for segmentation and visualization of curvi-
linear structures in medical images. *Cvrmed-Mrcas'97 1205*, 213–222.

[132] Schild, H. H. (1994). *MRI Made Easy (... well almost)*.

[133] Schlichting, H. and K. Gersten. *Boundary Layer Theory* (8th rev. and enl.
ed.). New York: Springer.

[134] Schubert, R. and M. J. Mulvany (1999). The myogenic response: established
facts and attractive hypotheses. *Clinical Science 96*(4), 313–326.

[135] Schumann, P., O. Touzani, A. R. Young, J. C. Baron, R. Morello, and E. T.
MacKenzie (1998). Evaluation of the ratio of cerebral blood flow to cerebral blood
volume as an index of local cerebral perfusion pressure. *Brain 121*, 1369–1379.

[136] Siegel, G. J. (2006). Basic neurochemistry : molecular, cellular, and medical
aspects.

[137] Smits, P., S. B. Williams, D. E. Lipson, P. Banitt, G. A. Rongen, and M. A.
Creager (1995). Endothelial release of nitric-oxide contributes to the vasodilator
effect of adenosine in humans. *Circulation 92*(8), 2135–2141.

[138] Stefanadis, C., J. Dernellis, E. Tsiamis, L. Diamantopoulos, A. Michaelides,
and P. Toutouzas (2000). Assessment of aortic line of elasticity using polynomial
regression analysis. *Circulation 101*(15), 1819–1825.

[139] Stefanadis, C., C. Stratos, C. Vlachopoulos, S. Marakas, H. Boudoulas,
I. Kallikazaros, E. Tsiamis, K. Toutouzas, L. Sioros, and P. Toutouzas (1995).
Pressure-diameter relation of the human aorta - a new method of determination
by the application of a special ultrasonic dimension catheter. *Circulation 92*(8),
2210–2219.

[140] Stefani, M. A., F. L. Schneider, A. C. H. Marrone, A. G. Severino, A. P.
Jackowski, and M. C. Wallace (2000). Anatomic variations of anterior cerebral
artery cortical branches. *Clinical Anatomy 13*(4), 231–236.

[141] Steinback, C. D., D. D. O'Leary, J. Bakker, A. D. Cechetto, H. M. Ladak,
and J. K. Shoemaker (2005). Carotid distensibility, baroreflex sensitivity, and
orthostatic stress. *Journal of Applied Physiology 99*(1), 64–70.

[142] Strandgaard, S. (1976). Autoregulation of cerebral blood-flow in hypertensive patients - modifying influence of prolonged antihypertensive treatment on tolerance to acute, drug-induced hypotension. *Circulation 53*(4), 720–727.

[143] Strzelczyk, J. (2003). The essential physics of medical imaging, 2nd edition. *Health Physics 85*(2), 242–242.

[144] Thoman, W. J., D. Gravenstein, J. van der Aa, and S. Lampotang (1999). Autoregulation in a simulator-based educational model of intracranial physiology. *Journal of Clinical Monitoring and Computing 15*, 481–491.

[145] Thoman, W. J., S. Lampotang, D. Gravenstein, and J. van der Aa (1998a). A computer model of intracranial dynamics integrated to a full-scale patient simulator. *Computers and Biomedical Research 31*(1), 32–46.

[146] Thoman, W. J., S. Lampotang, D. Gravenstein, and J. van der Aa (1998b). A computer model of intracranial dynamics integrated to a full-scale patient simulator. *Computers and Biomedical Research 31*(1), 32–46.

[147] Tian, R., P. Vogel, N. A. Lassen, M. J. Mulvany, F. Andreasen, and C. Aalkjaer (1995). Role of extracellular and intracellular acidosis for hypercapnia-induced inhibition of tension of isolated rat cerebral-arteries. *Circulation Research 76*(2), 269–275.

[148] Toda, N. and T. Okamura (2003). The pharmacology of nitric oxide in the peripheral nervous system of blood vessels. *Pharmacological Reviews 55*(2), 271–324.

[149] Tortora, G. J. and S. R. Grabowski (2003). *Principles of anatomy and physiology* (10th ed.). New Jersey: John Wiley and Sons, Inc. Tortora, Gerard J. and Grabowski, Sandra R.

[150] Treece, G. M., R. W. Prager, and A. H. Gee (1999). Regularised marching tetrahedra: improved iso-surface extraction. *Computers and Graphics 23*, 583–598.

[151] Ursino, M. (1988a). A mathematical study of human intracranial hydrodynamics - 1 the cerebrospinal-fluid pulse pressure. *Annals of Biomedical Engineering 16*(4), 379–401.

[152] Ursino, M. (1988b). A mathematical study of human intracranial hydrodynamics - 2 simulation of clinical-tests. *Annals of Biomedical Engineering 16*(4),
403–416.

[153] Ursino, M. (1991a). A mathematical-model of overall cerebral blood-flow regulation in the rat. *Ieee Transactions on Biomedical Engineering 38*(8), 795–807.

[154] Ursino, M. (1991b). A mathematical-model of overall cerebral blood-flow regulation in the rat. *Ieee Transactions on Biomedical Engineering 38*(8), 795–807.

[155] Ursino, M. and P. Digiammarco (1991). A mathematical-model of the relationship between cerebral blood-volume and intracranial-pressure changes - the
generation of plateau waves. *Annals of Biomedical Engineering 19*(1), 15–42.

[156] Ursino, M., P. Digiammarco, and E. Belardinelli (1989a). A mathematical-
model of cerebral blood-flow chemical-regulation - 1 diffusion-processes. *Ieee
Transactions on Biomedical Engineering 36*(2), 183–191.

[157] Ursino, M., P. Digiammarco, and E. Belardinelli (1989b). A mathematical-
model of cerebral blood-flow chemical-regulation - 2 reactivity of cerebral vascular
bed. *Ieee Transactions on Biomedical Engineering 36*(2), 192–201.

[158] Ursino, M. and M. Giulioni (2003). Quantitative assessment of cerebral autoregulation from transcranial doppler pulsatility: a computer simulation study.
Med Eng Phys 25(8), 655–66.

[159] Ursino, M. and C. A. Lodi (1997). A simple mathematical model of the
interaction between intracranial pressure and cerebral hemodynamics. *Journal of
Applied Physiology 82*(4), 1256–1269.

[160] Ursino, M., C. A. Lodi, and G. Russo (2000). Cerebral hemodynamic response
to co2 tests in patients with internal carotid artery occlusion - modeling study
and in vivo validation. *Journal of Vascular Research 37*(2), 123–133.

[161] Ursino, M. and E. Magosso (2000a). Acute cardiovascular response to isocap-
nic hypoxia - i a mathematical model. *American Journal of Physiology-Heart and
Circulatory Physiology 279*(1), H149–H165.

[162] Ursino, M. and E. Magosso (2000b). Acute cardiovascular response to iso-
capnic hypoxia - ii model validation. *American Journal of Physiology-Heart and
Circulatory Physiology 279*(1), H166–H175.

[163] Ursino, M., A. Ter Minassian, C. A. Lodi, and L. Beydon (2000). Cerebral hemodynamics during arterial and co2 pressure changes - in vivo prediction by a mathematical model. *American Journal of Physiology-Heart and Circulatory Physiology 279*(5), H2439–H2455.

[164] Viedma, A., C. JimenezOrtiz, and V. Marco (1997). Extended willis circle model to explain clinical observations in periorbital arterial flow. *Journal of Biomechanics 30*(3), 265–272.

[165] Walters, F. J. M. (1998). Intracranial pressure and cerebral blood flow. *Update in Anaesthesia 8*, 1–4.

[166] Wilhelms, J. and A. van Gelder (1990). Topological considerations in isosurface generation. *ACM Computer Graphics 24*(5), 79.

[167] Yeon, D. S., J. S. Kim, D. S. Ahn, S. C. Kwon, B. S. Kang, K. G. Morgan, and Y. H. Lee (2002). Role of protein kinase c- or rhoa-induced ca2+ sensitization in stretch-induced myogenic tone. *Cardiovascular Research 53*(2), 431–438.

[168] Zagzoule, M. and J. P. Marcvergnes (1986). A global mathematical-model of the cerebral-circulation in man. *Journal of Biomechanics 19*(12), 1015–1022.

[169] Zamir, M. (2000). *The physics of pulsatile flow.* Biological physics series. New York: AIP Press Springer.

[170] Zulch, K. J. and A. Agnoli (1971). *Cerebral circulation and stroke.* New York,: Springer. Editor: K. J. Zulch. With contributions by A. Agnoli [and others] illus. 26 cm.

www.ingramcontent.com/pod-product-compliance
Lightning Source LLC
LaVergne TN
LVHW022300060326
832902LV00020B/3182